Introduction to Bioplastics Engineering

Plastics Design Library

PLASTICS DESIGN LIBRARY (PDL)
PDL HANDBOOK SERIES
Series Editor: Sina Ebnesajjad, PhD (sina@FluoroConsultants.com)
President, FluoroConsultants Group, LLC
Chadds Ford, PA, USA
www.FluoroConsultants.com

The **PDL Handbook Series** is aimed at a wide range of engineers and other professionals working in the plastics industry, and related sectors using plastics and adhesives.

PDL is a series of data books, reference works and practical guides covering plastics engineering, applications, processing, and manufacturing, and applied aspects of polymer science, elastomers, and adhesives.

Recent Titles in the Series

Biopolymers: Processing and Products, Michael Niaounakis (ISBN: 9780323266987)
Biopolymers: Reuse, Recycling, and Disposal, Michael Niaounakis (ISBN: 9781455731459)
Carbon Nanotube Reinforced Composites, Marcio Loos (ISBN: 9781455731954)
Extrusion, 2e, John Wagner & Eldridge Mount (ISBN: 9781437734812)
Fluoroplastics, Volume 1, 2e, Sina Ebnesajjad (ISBN: 9781455731992)
Handbook of Biopolymers and Biodegradable Plastics, Sina Ebnesajjad (ISBN: 9781455728343)
Handbook of Molded Part Shrinkage and Warpage, Jerry Fischer (ISBN: 9781455725977)
Handbook of Polymer Applications in Medicine and Medical Devices, Kayvon Modjarrad & Sina Ebnesajjad (ISBN: 9780323228053)
Handbook of Thermoplastic Elastomers, Jiri G Drobny (ISBN: 9780323221368)
Handbook of Thermoset Plastics, 2e, Hanna Dodiuk & Sidney Goodman (ISBN: 9781455731077)
High Performance Polymers, 2e, Johannes Karl Fink (ISBN: 9780323312226)
Introduction to Fluoropolymers, Sina Ebnesajjad (ISBN: 9781455774425)
Ionizing Radiation and Polymers, Jiri G Drobny (ISBN: 9781455778812)
Manufacturing Flexible Packaging, Thomas Dunn (ISBN: 9780323264365)
Plastic Films in Food Packaging, Sina Ebnesajjad (ISBN: 9781455731121)
Plastics in Medical Devices, 2e, Vinny Sastri (ISBN: 9781455732012)
Polylactic Acid, Rahmat et. al. (ISBN: 9781437744590)
Polyvinyl Fluoride, Sina Ebnesajjad (ISBN: 9781455778850)
Reactive Polymers, 2e, Johannes Karl Fink (ISBN: 9781455731497)
The Effect of Creep and Other Time Related Factors on Plastics and Elastomers, 3e, Laurence McKeen (ISBN: 9780323353137)
The Effect of Long Term Thermal Exposure on Plastics and Elastomers, Laurence McKeen (ISBN: 9780323221085)
The Effect of Sterilization on Plastics and Elastomers, 3e, Laurence McKeen (ISBN: 9781455725984)
The Effect of Temperature and Other Factors on Plastics and Elastomers, 3e, Laurence McKeen (ISBN: 9780323310161)
The Effect of UV Light and Weather on Plastics and Elastomers, 3e, Laurence McKeen (ISBN: 9781455728510)
Thermoforming of Single and Multilayer Laminates, Ali Ashter (ISBN: 9781455731725)
Thermoplastics and Thermoplastic Composites, 2e, Michel Biron (ISBN: 9781455778980)
Thermosets and Composites, 2e, Michel Biron (ISBN: 9781455731244)

To submit a new book proposal for the series, or place an order, please contact David Jackson, Acquisitions Editor
david.jackson@elsevier.com

Introduction to Bioplastics Engineering

Syed Ali Ashter

R&D Endovascular
Maquet Getinge Group
Merrimack, NH, USA

ELSEVIER

AMSTERDAM • BOSTON • HEIDELBERG • LONDON
NEW YORK • OXFORD • PARIS • SAN DIEGO
SAN FRANCISCO • SINGAPORE • SYDNEY • TOKYO
William Andrew is an imprint of Elsevier

William Andrew is an imprint of Elsevier
The Boulevard, Langford Lane, Kidlington, Oxford, OX5 1GB, UK
50 Hampshire Street, 5th Floor, Cambridge, MA 02139, USA

British Library Cataloguing-in-Publication Data

A catalogue record for this book is available from the British Library.

Library of Congress Cataloging-in-Publication Data

A catalog record for this book is available from the Library of Congress.

ISBN: 978-0-323-39396-6

For Information on all William Andrew publications
visit our website at https://www.elsevier.com/

Publisher: Matthew Deans
Acquisition Editor: David Jackson
Editorial Project Manager: Zara Preston
Production Project Manager: Nicky Carter
Designer: Mark Rogers

In memory of my father, (Late) Professor Zaheer Haider

Contents

Preface xi
Acknowledgments xiii

1 Introduction 1
1.1 Background 1
1.2 Understanding Global Markets 5
1.3 Current Material Trends 6
1.4 Current Market Trends 10
1.5 Barriers to Biopolymer Commercialization 14
References 15

2 Overview of Biodegradable Polymers 19
2.1 Introduction 19
2.2 Definitions 22
2.2.1 Biodegradable 23
2.2.2 Biopolymer 24
2.2.3 Bio-based 26
2.2.4 Oxo-degradable 27
References 28

3 Mechanisms of Polymer Degradation 31
3.1 Introduction 31
3.2 Polyesters 33
3.3 Polyamides 38
3.3.1 Thermal Degradation 39
3.3.2 Photo-Oxidation 41
3.3.3 Thermal Oxidation 43
3.4 Polycarbonates 44
3.5 Polyurethanes 51
3.6 Polyacetals 53
3.6.1 Thermal Oxidation 53
3.6.2 Photo-Oxidation 54
3.6.3 Hydrolysis 55
References 56

4 Fundamentals on Biodegradability **61**
4.1 Introduction 61
4.1.1 What Is Biodegradability? 61
4.1.2 Biodegradability Standards 63
4.1.3 Biodegradability Testing 67
4.2 Measurement of Biodegradability of Polymers 70
4.2.1 Biodegradability Measuring Devices 73
4.3 Factors Affecting Biodegradability 77
References 78

5 Types of Biodegradable Polymers **81**
5.1 Introduction 81
5.2 Bio-Based Polymers 84
5.2.1 Starch-Based Polymers 86
5.2.2 Cellulose-Based Polymers 92
5.2.3 Chitin and Chitosan 103
5.2.4 Bacterial Polyesters 104
5.3 Synthetic Biodegradable Polymers 112
5.3.1 Poly(lactic acid) 114
5.3.2 Poly(ε-caprolactone) 118
5.3.3 Poly(glycolide) or Poly(glycolic acid) 121
5.3.4 Poly(*p*-dioxanone) 123
5.3.5 Poly(lactide-*co*-glycolide) 125
5.4 Bio-Derived Polyethylene 127
5.5 Polymers from Bio-Based Monomers 132
5.5.1 Lactic Acid and Poly(lactic acid) 133
5.5.2 1,3 Propanediol and DuPont's Sorona 134
5.5.3 Succinic Acid 136
5.5.4 Bio-Based Unsaturated Polyester 136
5.6 Bio-Based Materials for Paper/Substrate Coatings 139
References 141

6 Additives and Modifiers for Biopolymers **153**
6.1 Introduction 153
6.2 Types of Additives and Modifiers 154
6.2.1 Type 1 Additives 155
6.2.2 Type 2 Additives 155
6.2.3 Type 3 Additives 156
6.3 Additives and Modifiers for Bioolefins 156

6.4 Additives and Modifiers for Biopolymers 160
6.4.1 Toughness 160
6.4.2 Processability 163
6.4.3 Heat Resistance 165
6.4.4 Foamability 167
6.4.5 Mechanical Properties 167
6.5 Effect on Properties 172
6.5.1 Toughness 172
6.5.2 Processability 173
6.5.3 Heat Resistance 173
6.5.4 Foamability 174
6.5.5 Mechanical Properties 175
References 175

7 Processing Biodegradable Polymers 179
7.1 Introduction 179
7.2 Injection Molding 180
7.2.1 NatureWorks Ingeo Series 182
7.2.2 Novamont Mater-Bi 184
7.2.3 DuPont Biomax TPS 184
7.2.4 Metabolix Mirel 184
7.2.5 TianAn Biologic 186
7.2.6 Ecovio IS 1335 186
7.3 Compression Molding 187
7.4 Film Casting 189
7.5 Blow Molding 192
7.5.1 Stretch Blow Molding 194
7.6 Blown Film Extrusion 196
7.7 Thermoforming 197
7.8 Fiber Spinning 199
7.8.1 Melt Spinning 200
7.8.2 Electrospinning 202
References 207

8 Extrusion of Biopolymers 211
8.1 Introduction 211
8.2 Conventional Extrusion of Biopolymers 213
8.2.1 LSLF Extrusion 215
8.3 Starch Extrusion 216

8.4 Extrusion Screw Designs 217
8.4.1 Turbo-Screw Design 218
8.4.2 Pulverization Screw Design 220
8.4.3 Reactive Extrusion Screw Design 221
References 223

9 Commercial Applications of Bioplastics 227
9.1 Introduction 227
9.2 Packaging 228
9.3 Bags 230
9.4 Disposable Housewares 231
9.5 Agriculture and Horticulture 234
9.6 Medical Devices 235
9.7 Consumer Electronics 240
9.8 Automotive 240
References 246

10 New Developments 251
10.1 Introduction 251
10.2 Materials 251
10.2.1 Expanded PLA (EPLA) Bead and Foams 255
10.2.2 Ecodear Nanoalloys 258
10.2.3 Ecovio L Foam 259
10.2.4 Ingeo 3D Series 259
10.2.5 Expansorb 261
10.2.6 LACTEL Absorbable Polymers 263
10.3 Processing 264
10.4 Applications 266
References 272

Index 275

Preface

Introduction to Bioplastics Engineering is a practical, user-friendly reference for plastics engineers working with biopolymers and biodegradable plastics. This book provides plastics engineers and researchers with a fundamental, practical understanding of the differences between bioplastics and biodegradable polymers and guidance on the different methods used to process bioplastics. This book also covers additives and modifiers for biopolymers and their effect on properties. Examples are included of commercial applications of bioplastics, as well as new bioplastics being developed and future trends in the industry.

Introduction to Bioplastics Engineering consists of 10 chapters that will enable engineers, researchers, technicians, and students a sound understanding on bioplastic background and its market. Chapter 1: Introduction provides discussion on global bioplastic market, current material and market trends, and limitation toward biopolymer commercialization. Chapter 2: Overview of Biodegradable Polymers focuses on defining biodegradable, biopolymer, bio-based, and oxo-degradable terminologies.

Chapter 3: Mechanisms of Polymer Degradation reviews different mechanisms of polymer degradation. In this chapter, degradation mechanism of five different engineering polymers, some of which are commercially available and some that has potential for the future. Chapter 4: Fundamentals on Biodegradability will discuss fundamentals on biodegradability, testing standards, and ways to measure biodegradability. Chapter 5: Types of Biodegradable Polymers will review different types of bio-based and synthetic biopolymers. It will also highlight on polymers derived from monomers as well as bio-derived polyethylene.

Chapter 6: Additives and Modifiers for Biopolymers talks about different types of additives and modifiers for biopolymers and study their effects on properties. Chapter 7: Processing Biodegradable Polymers includes discussion on different ways to process biodegradable polymers. Processing of some of the commercial biodegradable polymers will also be discussed. Chapter 8: Extrusion of Biopolymers reviews extrusion of biopolymers. In this chapter, conventional extrusion process for biopolymer processing, starch extrusion, and different extrusion screw designs will be discussed.

Chapter 9: Commercial Applications of Bioplastics discusses different areas of commercial applications of biodegradable plastics such as packaging, bags and sacks, disposable housewares, agriculture and horticulture, medical devices, consumer electronics, and automotive. New developments of bioplastic in material, processing, and applications are reviewed in Chapter 10: New Developments.

Syed Ali Ashter
R&D Endovascular
Maquet Getinge Group
Merrimack, NH, USA

Acknowledgments

I would like to show my sincere appreciation to Professor Stephen Burke Driscoll from Plastics Engineering Department, University of Massachusetts Lowell for providing valuable inputs and guidance. I would also like to thank Dr Sina Ebnesajjad, Plastics Design Library Handbook Series editor for his support and constant mentoring during the book-writing process. I would also like to thank my family especially my wife, Tahira, and my two kids, Zayn and Noor for their unconditional support. This work would have never been completed otherwise.

1 Introduction

1.1 Background

Over the last several years, production of polymers from renewable resources has shown significant growth. Some of the plastics produced from renewable resources such as vegetable oil, corn, and pea starch have been synthesized by microbes and are known as bioplastics. Its development is driven by current demands to replace fossil fuel–based polymers. Limitation in fossil fuel resources, price volatility, impact on the environment, and waste disposal problems are some of the main reasons for this shift toward bio-based plastics.

The use of natural polymers is not a new idea. Fig. 1.1a–c shows natural resins like amber, shellac, and gutta percha that were used during Roman Times and the Middle Ages [1–3]. Native Americans were developing and refining techniques for making ladles and spoons from animal horns long before there was any European contact. Initial development work began in the 1920s when Ford Motor Co. began experimenting with soybeans in the automobiles. However, it all began in the 1940s when Ford Motor Co. gave a go-ahead to produce plastic parts from soybeans to support the idea of sustainability [4].

Bioplastics are broadly classified as bio-based and/or biodegradable. When the focus of the material is on the origin of the carbon building blocks and not by where it goes at the end of its product life, it is termed as bio-based. It is important to understand that all bio-based materials are not often characterized as biodegradable, and similarly, not all biodegradable materials are bio-based. Material is considered biodegradable when materials are broken down under the influence of microbes and right conditions and use them as food source. When a complete microbial assimilation of the fragmented food source happens within 180 days in a compost environment, it is considered as compostable. Fig. 1.2 shows pictorial difference between two branches of bioplastics—bio-based and biodegradable, respectively [5,6].

American Society for Testing and Materials (ASTM) developed a standardized test method, ASTM D6866 to determine bio-based content. Originally, developed for the US Department of Agriculture (USDA)

Introduction to Bioplastics Engineering.
DOI: http://dx.doi.org/10.1016/B978-0-323-39396-6.00001-4

Figure 1.1 Natural resins used during Roman Times and Middle Ages: (a) Amber, (b) Shellac, and (c) Gutta Percha [1–3].

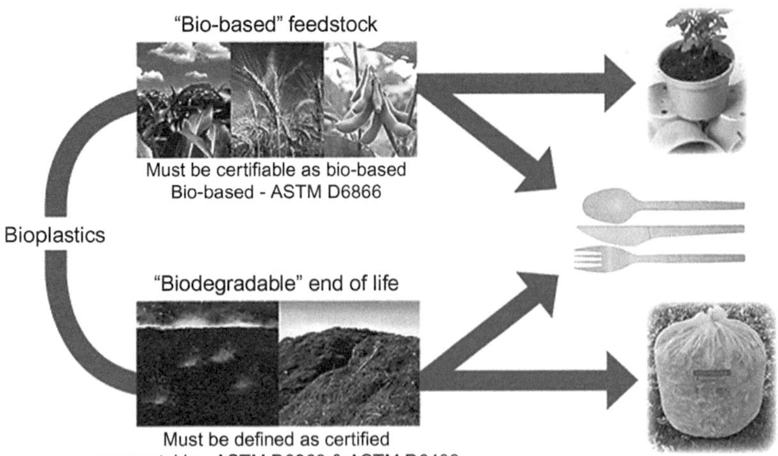

Figure 1.2 Differentiating between biodegradable and bio-based [6].

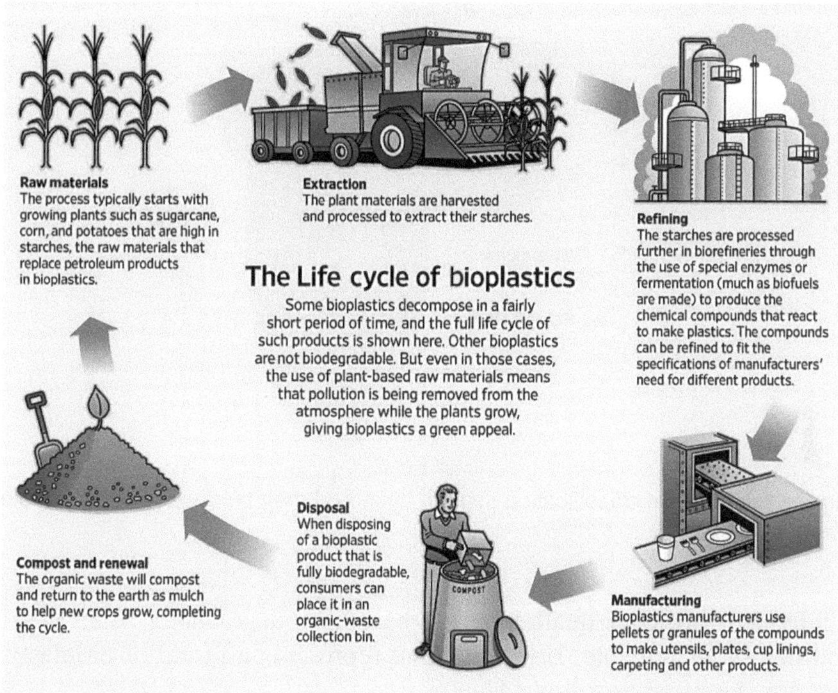

Figure 1.3 The life cycle of bioplastics [9]. *CTC Clean Tech Consulting GmbH; WSJ reporting.*

Bio-Preferred program, this test method uses radiocarbon dating to determine bio-based contents of materials. Standardized test methods, ASTM D6868 and ASTM D6400, were developed to set specification for compostable plastics [5,7,8].

Fig. 1.3 provides an illustration on the life cycle of bioplastics. It all starts with growing plants such as sugarcane and corn that are high in starches. The plants are then harvested and processed to extract their starches. The extracted starches are then refined and fermented using special enzymes producing chemical compound that react to make plastics. Plastics in the form of pellets are used to manufacture products. The product at this stage is fully biodegradable. After its full use, the product is then placed in an organic waste container, which starts the last stage of the cycle [9,10].

Fig. 1.4 shows material coordinate system to classify type based on their bio-based content and biodegradability. The coordinate system is

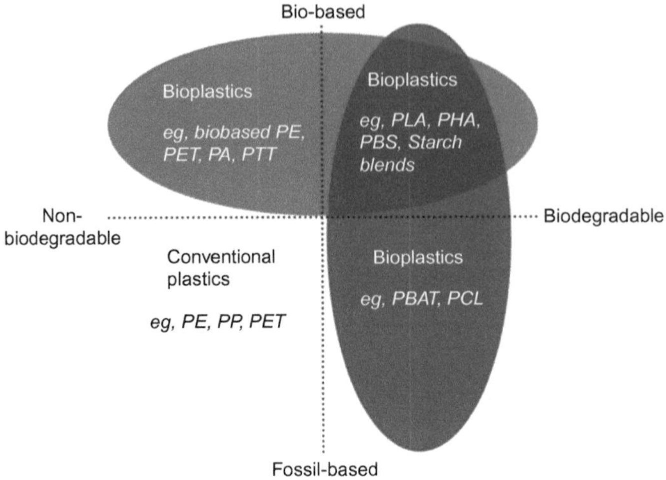

Figure 1.4 Material classification system based on their biodegradability and bio-based content [11].

subdivided into four quadrants: bio-based, biodegradable, fossil-based, and non-biodegradable. Each quadrant represents a group of bioplastics that are classified as follows [11]:

- Group 1 — Bioplastics that are bio-based or partly bio-based non-biodegradable such as bio-based PE, PET, PA, and PTT.
- Group 2 — Bioplastics that are bio-based and biodegradable, such as PLA, PHA, PBS, and starch blends.
- Group 3 — Bioplastics that are fossil-based and non-biodegradable such as conventional PE, PP, and PET.
- Group 4 — Bioplastics that are fossil-based and biodegradable such as PBAT and PCL.

To have sustainable production and consumption, bioplastics have some added benefits. Some of the several advantages are as follows:

- Increase in efficiency
- Renewable resource that can be cultivated annually
- Reduction in carbon footprint

Currently, bioplastics is in infancy and going through growth phase. There are lots of expectations pinned on bioplastics and many aspects have to be evaluated to make the process commercially viable.

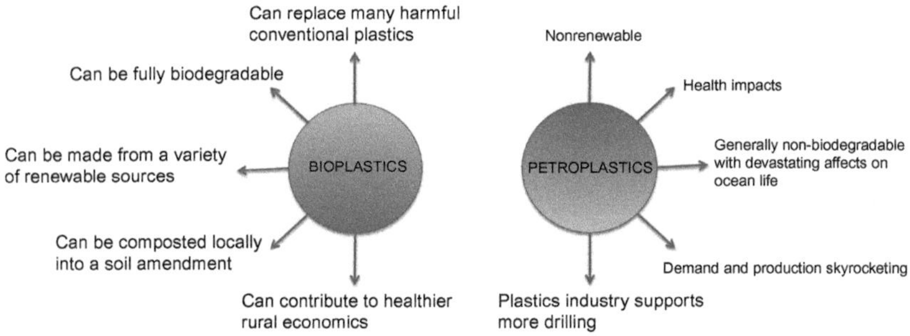

Figure 1.5 Differences between bioplastics and petroplastics [12].

Cost feasibility is the most important of all however, other factors such as concerns about genetically modified organisms, sustainably grown biomass, composting programs and infrastructure, lack of adequate labeling, and concern over contamination of recycling systems have to be thoroughly understood. Despite all these points, bioplastics have many merits over the petroplastics as shown in Fig. 1.5 [12].

1.2 Understanding Global Markets

Bioplastics are a type of plastic that can be made from natural resources such as vegetable oils and starches. Since bioplastics are plant-based products, the consumption of petroleum for the production of plastic is expected to decrease by 15–20% by 2025. By 2025, Asia and Europe will have the largest share of bioplastics market. Asia will account for 32% while Europe at 31% of the total market followed by the United States at 28%. Currently, bioplastics market growth is at 10% annually covering approximately 10–15% of the total plastics market. This number would increase to 25–30% by 2020 [13].

Nova Institute has done a global survey of 247 corporations covering almost all major bio-based plastics. Based on the survey, they estimate that the bioplastic production capacity will increase to nearly 12 million tons by 2020. This would increase the bio-based share from 1.5% in 2011 to 3% in 2020. They estimate that most investment in bio-based capacities will take place in Asia and South America because of better access to the feed source. Fig. 1.6 shows an overview of the change in global bioplastics productions by regions [14]. Europe's share will decrease from 20% to 14% and North America will see a decrease from 15% to 13%. Asia's bioplastics production share will see an increase from 52% to 55%, whereas that of South America will see a jump from 13% to 18%.

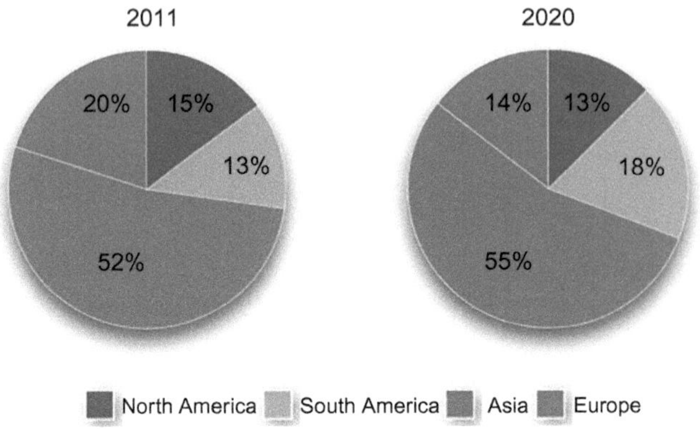

Figure 1.6 Overview of the change in global bioplastics productions by regions [14].

European Bioplastics Association estimated that 58% of 1.16 million metric tons global bioplastics capacity in 2011 was bio-based. Global bioplastics capacity will see almost a fivefold increase from 2011 to 2016. The composition of bioplastics production capacity is also expected to change significantly from 58% bio-based/non-biodegradable in 2012 to 87% by 2016. Some of the major factors driving the bioplastic market are high consumer acceptance, high fossil fuel prices, increase in the dependence on fossil fuels, and the need for more ecofriendly products. According to a study by Helmut Kaiser Consultancy, less than 3% of all waste plastic worldwide gets recycled, compared with recycling rates of 30% for paper, 35% for metals, and 18% for glass. Currently, the demand for bioplastics is increasing due to its renewability and availability of raw material, advanced functionality and technical properties, and the recycling option that they present. [13].

Some of the major players competing in this market are Cargill's NatureWorks, DuPont, Braskem (BAK), Toray Industries, Lanxess AG, Bayer, BASF, and Eastman.

1.3 Current Material Trends

One of the most important advancements from the last few years has been the development of materials that are produced from monomer building blocks from the natural feedstock. These materials are termed as drop-ins and can directly replace conventional fossil-based plastics. These drop-ins open a new route to produce bimonomers that can easily fit into existing production cycle.

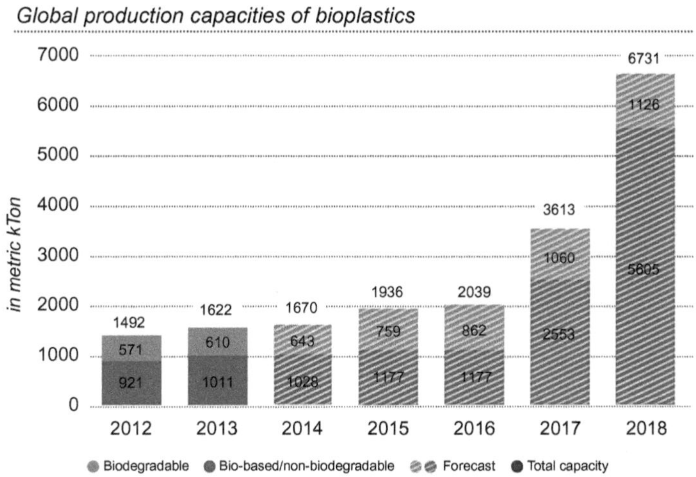

Figure 1.7 Global production capacities of bioplastics [15]. European Bioplastics, Institute for Bioplastics and Biocomposites, nova-Institute (2014). More information: www.bio-based.eu/markets and www.downloads.ifbb-hannover.de.

Among all drop-ins, partially bio-based PET leads the field. Bio-based PET accounts for approximately 40% of the global bioplastics production capacity. Fig. 1.7 shows global production capacities of bio-plastics. A 10-fold increase to 80% of total bioplastics production capacity is expected in 2018 to 5.6 million tons. Bio-based PE (polyethylene) follows bio-based PET, another drop-in material strongly driving bioplastics growth with more than 4% of the bio-based production capacity predicted for 2016. Fig. 1.8 shows global production capacities of bioplastics by material type. There are materials that have been or are being commercialized which include bio-based nylon, polypropylene, polystyrene, polycarbonate, polyvinyl chloride (PVC), and many other traditional plastics. Europe has the world's largest market for bio-plastics; however, production capacities of Asia and South America are seen to rapidly growing. [15,16].

Starch-based resins and polylactic acid (PLA) are projected to lead bioplastic products through 2017, combining to account for over 60% of demand. PLA demand will benefit from the development of resins and compounds with enhanced performance for more durable applications such as fibers, automotive parts, and electronic parts.

A significant shift from the first-generation feedstocks to second-generation feedstocks such as cellulosics will happen in coming years. Cellulosic feedstocks consist of crop residues, wood residues, yard

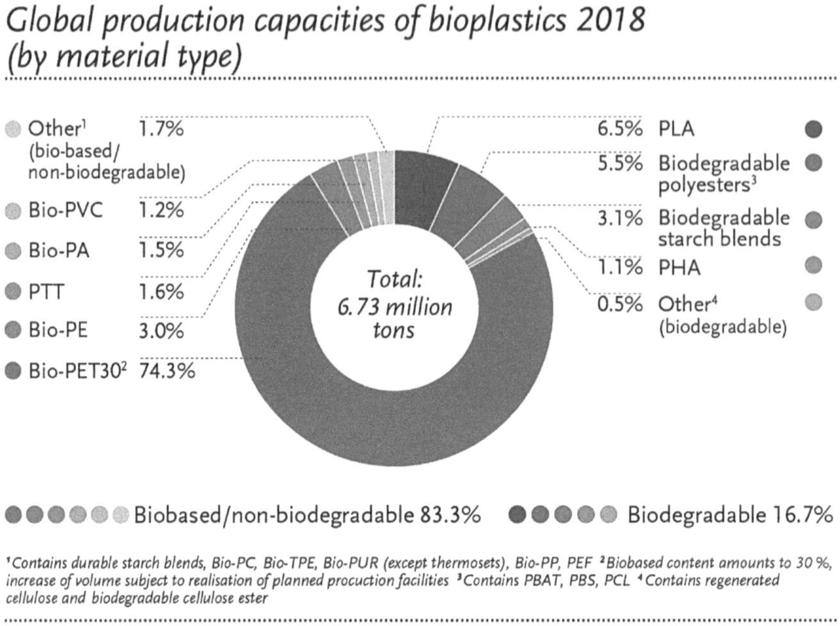

Figure 1.8 Global production capacities of bioplastics by material type [16]. European Bioplastics, Institute for Bioplastics and Biocomposites, nova-Institute (2014). More information: www.bio-based.eu/markets and www.downloads.ifbb-hannover.de.

waste, municipal solid waste, algae, or other biomass. They can be converted to sugars via various technologies, including enzymatic hydrolysis and biomass pretreatment. Cellulosic feedstocks currently produced are cellulose acetates and lignin-based polymers. However, in order to generate more cellulosic feedstocks, sophisticated biorefineries are needed that can perform the process steps needed to produce various bio-products as shown in Fig. 1.9. Once these are in place, a stream of nonfood crop–based fermentable sugars will become available for energy, chemicals, and polymers [17,18].

Bio-based additives and modifiers are another area which will see a strong development. Because of its added advantage and enlarged concerns that plasticizers used in PVC and Bisphenol-A in polycarbonate impose to the environment, there is a drive to find a health and environmentally friendly solutions. Bio-based additives are not only relevant for engineering durable biopolymers with enhanced performance properties but also used to develop an alternative to the conventional modifiers. Increasingly, bio-based formulations are also being used to

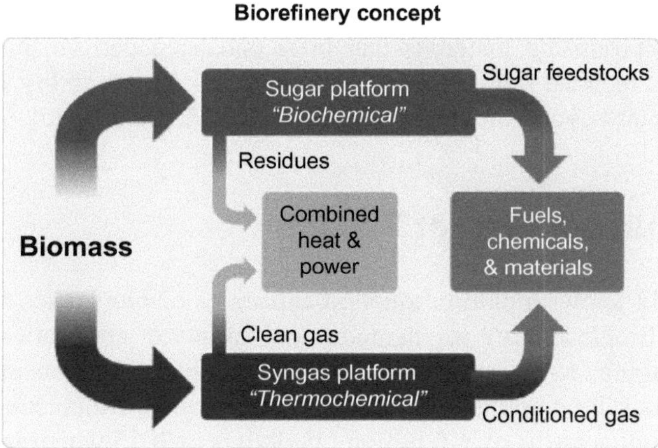

Figure 1.9 A concept of biorefinery [18].

Figure 1.10 PHA-based polymeric modifier developed by METABOLIX [20].

modify conventional materials, as these have been found to enhance the performance of these materials in various ways while at the same time improving their carbon footprint [19].

Metabolix Inc. has developed a series of PHA-based polymeric modifiers that demonstrate very good miscibility with PVC and improve its mechanical and environmental performance characteristics. This additive and modifier is sold under the trade name Mirel as shown in Fig. 1.10. Similarly, Mitsubishi Chemical produces a polycarbonate in which the Bisphenol-A is has been replaced by isosorbide, a biomonomer that can be safely used in food applications. Isosorbide-based copolyesters are extremely promising materials that offer enhanced performance properties. PLA, blended with PMMA, enhances the processability and other properties far beyond those of conventional acrylic resins [20].

1.4 Current Market Trends

Fig. 1.11 shows global production capacities of bioplastics by market segment. Bioplastics are segmented into a variety of applications including packaging, agriculture, food services, automotive, consumer electronics, household appliances, and consumer goods. Bioplastics markets will see high growths over 6 billion US dollars in 2015 which is expected to significantly grow to over 12.5 billion US dollar by 2020 [21]. Rigid packaging will outpace other markets by showing higher growth especially in food and beverage packing, catering products, and bags [16].

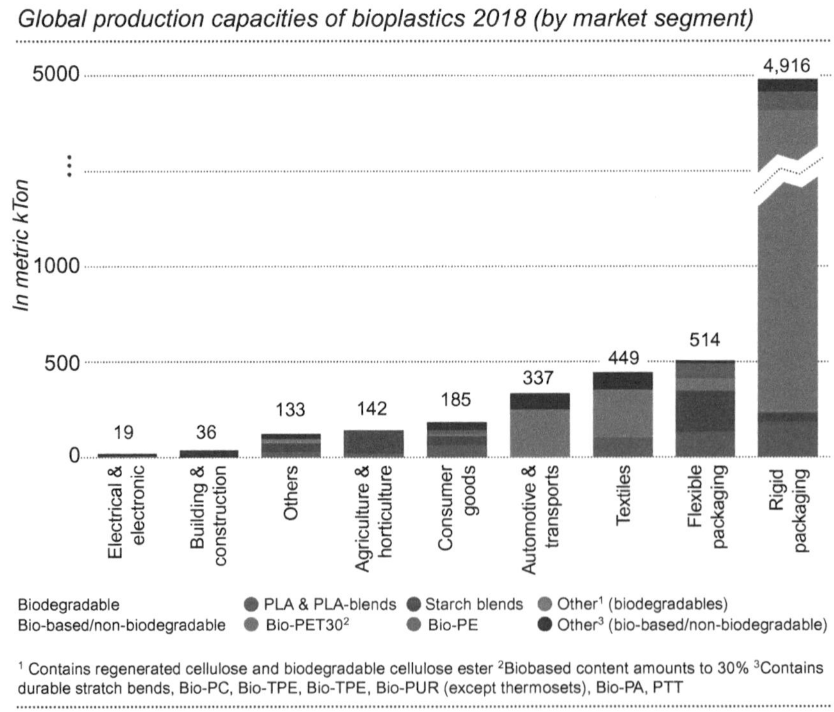

Figure 1.11 Global production capacities of bioplastics by 2018 [16]. European Bioplastics, Institute for Bioplastics and Biocomposites, nova-Institute (2014). More information: www.bio-based.eu/markets and www.downloads.ifbb-hannover.de.

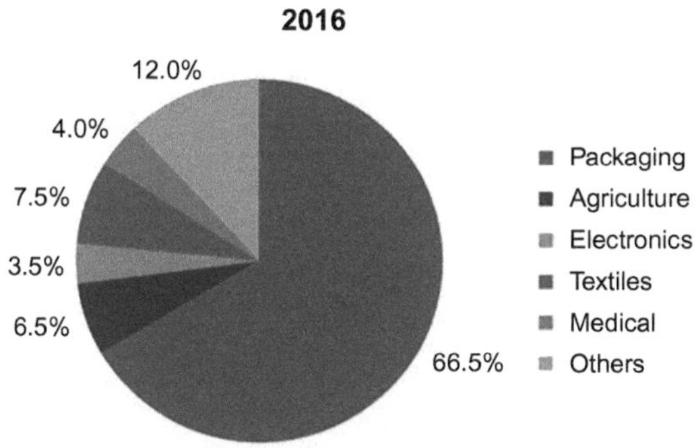

Figure 1.12 Bioplastics application by market share [22].

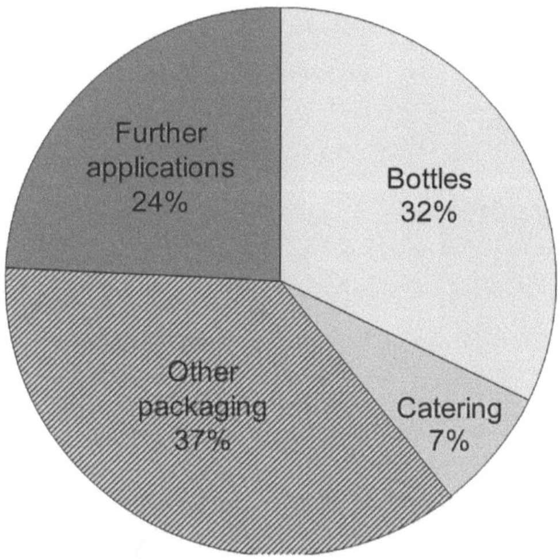

Figure 1.13 Packaging applications in bioplastics industry [23].

Fig. 1.12 shows bioplastics application by market share. The packaging industry by itself represents the biggest application field of these materials. It is estimated that 66% of the global bioplastic production is used in packaging applications and it will represent up to 80% of its global production in the coming years [22].

Fig. 1.13 shows packaging applications in the bioplastics industry. The increase in packaging is mainly due to the increase in PET-manufactured bottles made from renewable sources [23].

Packaging being the largest application, the most common bioplastics that is used for food packaging applications are plastics based on starch or cellulose, PLA, polyvinyl alcohol, aliphatic–aromatic copolyesters and polyethylene (PE) and polyethylene terephthalate (PET) partially or completely obtained from renewable sources (Bio-PE and/or Bio-PET). Some of the bioplastics applications by market share are shown in Fig. 1.14.

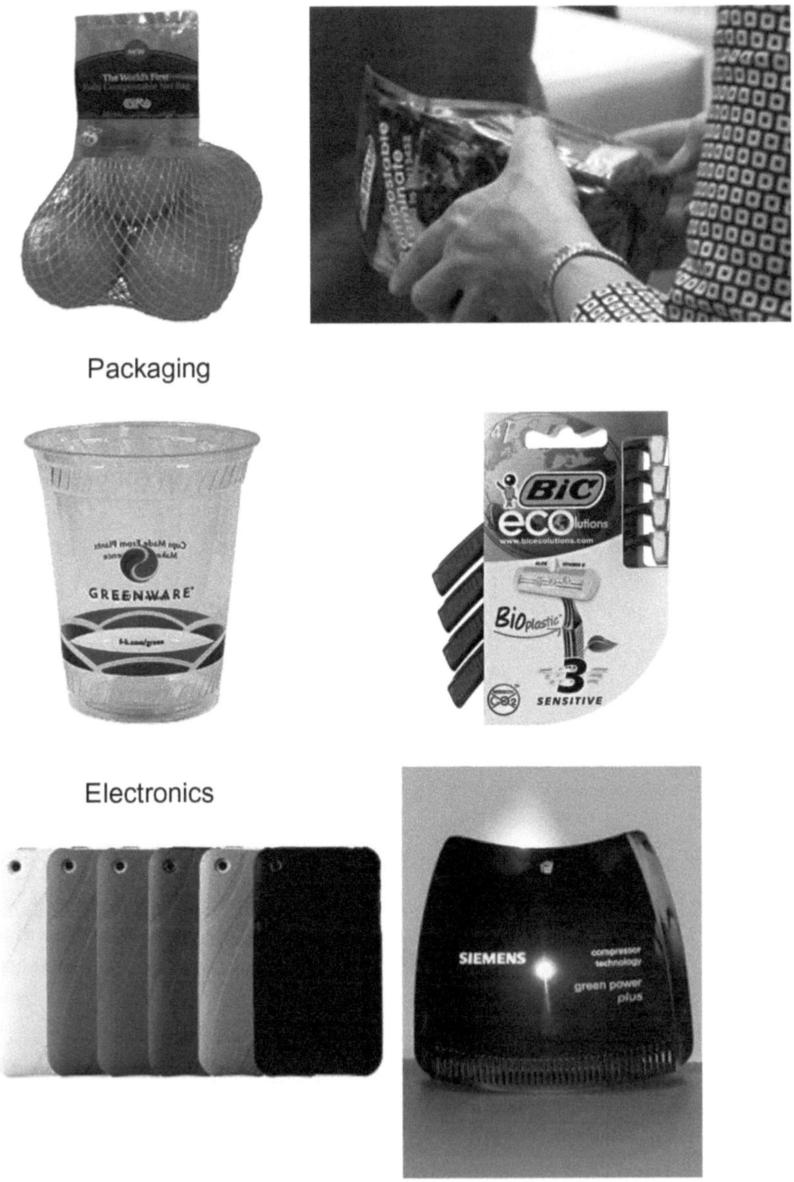

Figure 1.14 Bioplastics applications by market share [24–35].

Automotive

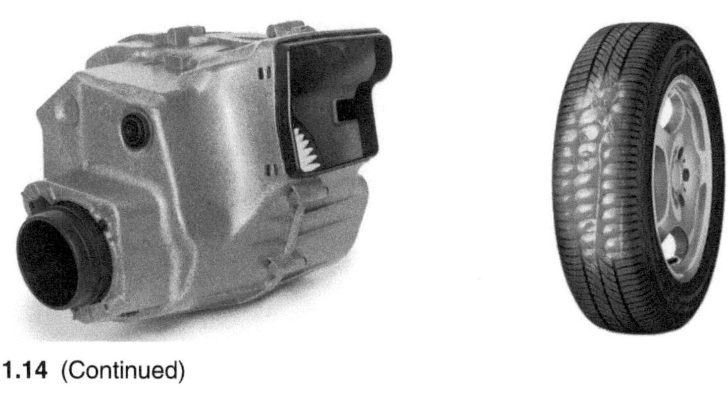

Figure 1.14 (Continued)

Figure 1.14 (Continued)

1.5 Barriers to Biopolymer Commercialization

Biopolymer development is still in its early stages. Due to the growing needs of the customers, there is a strong demand to develop products that are environment-friendly and sustainable. However, biopolymer innovations to market pose significant technological and application challenges. Constant increase in demand to manufacture biopolymers from the plants put pressure on agricultural crops as they have to satisfy the needs of the ever growing population. In addition, poor performance characteristics and high cost when compared against conventional polymers are the key drawback. High investment costs to process biopolymers also poses as a challenge [36].

The technology to develop number of bio-based non-biodegradable biopolymers like PA6, PA66, and Bio-PP is still in its infancy period, and these biopolymers are yet to mature at commercial stages. Plastic processing technologies that can efficiently process new generation of

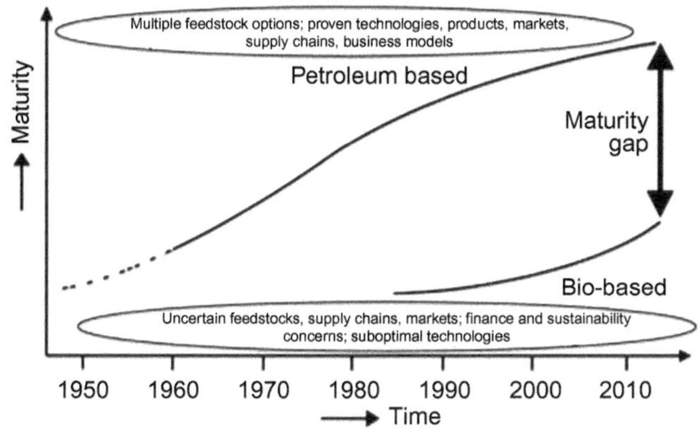

Figure 1.15 Major differences between conventional plastics and bioplastics [37].

biopolymers need to be developed and hence pose another major challenge. The product performance of biopolymers in the market is directly associated with their bio-based content in bioplastics. The area of challenge is in developing the right tool that can accurately measure bio-based content for bio-based biodegradable as well as non-biodegradable bioplastics [37].

Biopolymers face significant challenges in taking products to market faster and focus on quick solutions by integrating the product into the existing structures. The growth of biopolymers is hampered by their poor price competitiveness and lack of appropriate feedstocks. Fig. 1.15 shows how biopolymer market is doing against conventional polymer. Over the last decade, biopolymer market has seen large growth; however, there exist considerable gap for biopolymer market realization [37].

In comparison to conventional polymers, biopolymers are still in their early growth phase. They are yet to establish proven technology to drive growth through innovation, have a production platform to operate with reliability, establish a reliable supply-chain and establish customer base by marketing the product.

References

[1] Natural History of Amber, <http://www.houseofamber.com/eng/natural-history-of-amber/>.
[2] Shellac—Applications and Use in Art, <http://www.naturalpigments.com/art-supply-education/shellac-use-art>.
[3] Pilot Products, Inc., <http://www.pilotproducts.com/all-about-natural-rubber.html>.

[4] Benson Ford Research Center, <https://www.thehenryford.org/research/soybeancar.aspx>.

[5] Standard Specification for Labeling of Plastics Designed to Be Aerobically Composed in Municipal or Industrial Facilities, American Society for Testing Materials (ASTM), D6400-12, <http://www.astm.org/Standards/D6400.htm>.

[6] Natur Bag, <http://www.naturbag.com/frequently-asked-questions>.

[7] Standard Test Method for Determining the Biobased Content Of Solid, Liquid, and Gaseous Samples Using Radiocarbon Analysis, American Society for Testing Materials (ASTM), D6866-12, <http://www.astm.org/Standards/D6866.htm>.

[8] Standard Specification for Labeling of End Items That Incorporate Plastics and Polymers as Coating or Additives With Paper and Other Substrates Designed to Be Aerobically Composted in Municipal or Industrial Facilities, American Society for Testing Materials (ASTM), D6868-11, <http://www.astm.org/Standards/D6868.htm>.

[9] The Life Cycle of Bioplastics, <https://bootstrapcompost.wordpress.com/>.

[10] J.M. Bier, C.J.R. Verbeek, M.C. Lay, Life cycle assessments of bioplastics: applications and issues, Int. J. Environ. Cult. Econ. Soc. Sustainability 7 (4) (2011) 145–158 (CG Publisher).

[11] Bioplastics Material Coordinate System, European Bioplastics. <http://en.european-bioplastics.org/press/faq-bioplastics/>.

[12] Sustainable Plastics, <http://www.sustainableplastics.org/about>.

[13] Bioplastics Market Worldwide: Applications, Methods, Competition, Materials, Technologies, Development, Recycle, Renewable Energy, Production, Consumption, Helmut Kaiser Consultancy, <http://www.hkc22.com/bioplastics.html>.

[14] Bioplastics market to triple, European Biotechnology Magazine, <http://www.european-biotechnology-news.com/news/news/2013-01/bioplastics-production-to-triple.html>.

[15] Greener Package, <http://www.greenerpackage.com/bioplastics/data_predict_significant_growth_bioplastics_2018>.

[16] European Bioplastics, Institute for Bioplastics and Biocomposites, Nova Institute (2014).

[17] The future of industrial biorefineries, World Economic Forum, <http://www3.weforum.org/docs/WEF_FutureIndustrialBiorefineries_Report_2010.pdf>.

[18] Biofuels, Biochemical and Bioplastics, ViaSpace, <http://www.viaspace.com/biochemicals_bio_plastics.php>.

[19] N. Lardjane, N. Belhaneche-Bensemra, V. Massardier, Migration of new bio-based additives from rigid and plasticized PVC stabilized with epoxidized sunflower oil in soil, J. Polym. Res. 20 (2013) 209.

[20] K. Laird, Bioplastics in 2013: 5 trends to watch. <http://www.plasticstoday.com/articles/bioplastics-2013-5-trends-watch-122720121>.

[21] Bioplastics Market Forecast 2014−2024, Prospects for Bio-based/Non-Biodegradable Packaging, Consumer Electronics Products, Catering, Agriculture & Pharmaceuticals, 2014.
[22] Technical Article & Report on Plastics Industry, Biodegradable Plastics Demand to Grow 15% Annually to 2015, Plastemart, <http://www.plastemart.com/Plastic-Technical-Article.asp?LiteratureID=1958&Paper=biodegradable-plastics-demand-to-grow-15-percent-annually-to-2015>.
[23] C. Wagner, Bioplastics: Types, Applications, Toxicity and Regulation of Bioplastics Used in Food Contact Materials, Food Packaging Forum, Zurich, 2014.
[24] European Bioplastics, <http://en.european-bioplastics.org/market/applications/packaging/>.
[25] Biomass Packaging, <http://www.biomasspackaging.com/education/bioplastics/>/.
[26] BIC® Ecolutions, <http://ie.bicworld.com/shavers-desc/144-bicr-ecolutions-bics-first-shaver-with-a-bioplastic-handle.html>.
[27] European Bioplastics, <http://en.european-bioplastics.org/press/press-pictures/consumer-electronics/>.
[28] D. Smock, Siemens plastic replicates ABS with bioplastic and CO_2-off-take, The Molding Blog (2012).
[29] DSM in Automotive/Transportation, <http://www.dsm.com/markets/automotive/en_US/informationcenter-press/2015/01/2015-01-08-dsm-to-speak-on-bio-based-materials-at-bioplastics-conference.html>.
[30] A. Bredenberg, Do Bioplastics Have a Place in Automotive manufacturing, ThomasNet.com, <http://news.thomasnet.com/imt/2012/12/17/do-bioplastics-have-a-place-in-automotive-manufacturing>.
[31] Corbion, <http://www.corbion.com/bioplastics/markets/automotive>.
[32] A. Bredenberg, Bioplastics Can Find Diverse Applications in the World of Intermediates, ThomasNet.com, <http://news.thomasnet.com/imt/2013/01/07/bioplastics-can-find-diverse-applications-in-the-world-of-intermediates>.
[33] European Bioplastics, <http://en.european-bioplastics.org/market/applications/agriculturehorticulture/>.
[34] Interempresas.net, <https://www.interempresas.net/Plastico/Articulos/111494-Entrevista-Sandra-Pazes-Ventas-Iberia-Latinoamerica-David-Bertomeu-Tecnologia-Aplicacion.html>.
[35] Greener Package, <http://www.greenerpackage.com/compost_biodegrade/plantable_nursery_pot_biodegrades_soil>.
[36] R. Papalia, Turning plants into plastics: challenges and opportunities in bioplastics for consumer goods companies, Environ. Lead., <http://www.environmentalleader.com/2011/06/07/turning-plants-into-plastics-challenges-and-opportunities-in-bioplastics-for-consumer-goods-companies/>.
[37] R. Rajagopal, Sustainable Value Creation in the Fine and Speciality Chemicals Industry, Wiley Publishers, New Delhi, 2014.

2 Overview of Biodegradable Polymers

2.1 Introduction

Due to the increased consumption of fossil-based fuel, resource limitation, price fluctuation, and impact on the environment, there has been a considerable shift toward using biodegradable materials [1]. Polymer materials are comprised of repeating macromolecules known as mer units. Each mer unit is called a monomer, while multiple repeating units are known as polymers. Fig. 2.1 shows examples of the mer units for different polymer systems [2].

Plastics that are derived from fossil-based feedstocks resist degradation leading to discussions on how to dispose them. These discussions have been critical toward the development of biodegradable polymers. As shown in Fig. 2.2, biodegradable polymers come from various sources, from natural to synthetic polymers. Natural polymers are available in large quantities from renewable sources, while synthetic polymers are produced from nonrenewable petroleum resources [3,4]. Renewable resource feedstocks also include microbially grown polymers and those extracted from starch [5].

Fig. 2.3 shows a typical biodegradation process. In this process, organic molecules in the environment are broken down into simpler compounds by breaking bonds, either hydrolytically or by using bacteria, fungi, yeast, and their enzymes [6]. Under the ideal conditions of temperature, moisture, and oxygen, biodegradation process happens relatively fast [6]. Material usage and final mode of biodegradation are dependent on the composition and processing method employed. An integrated waste management system may be necessary in order to efficiently use, recycle, and dispose of biopolymer materials. Reduction in the consumption of sources, reuse of existing materials, and recycling of discarded materials must all be considered.

Biodegradable polymers are the solution to disposal problems commonly encountered with conventional polymers. In the case of biodegradable polymers, it is not necessary to recycle after the end of its useful life and they can be left in the environment to biodegrade.

Introduction to Bioplastics Engineering.
DOI: http://dx.doi.org/10.1016/B978-0-323-39396-6.00002-6

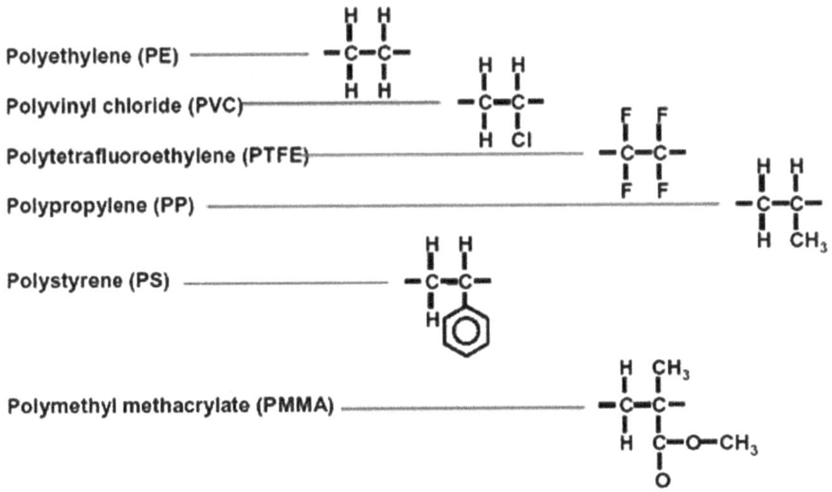

Figure 2.1 Examples of "mer units" for different polymer systems [2].

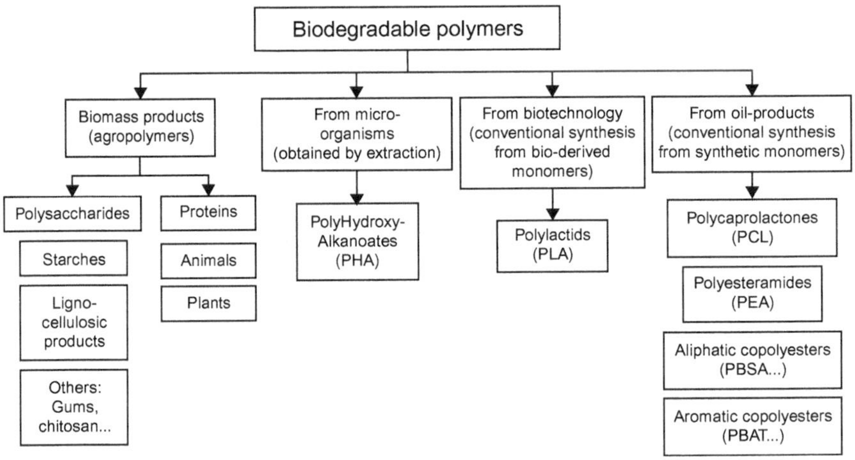

Figure 2.2 Classification of biodegradable polymers [4].

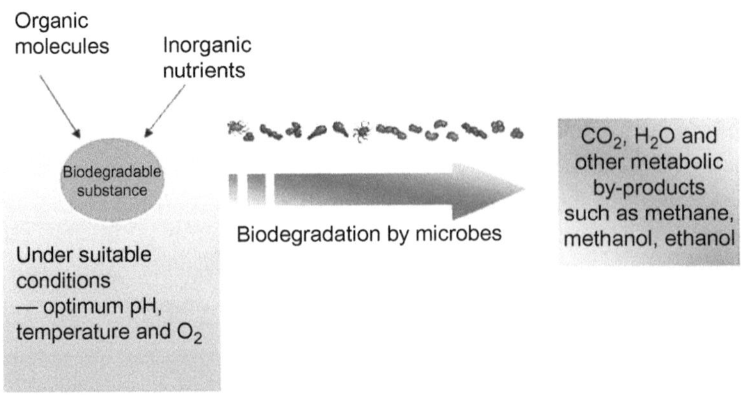

Figure 2.3 A typical biodegradation process [6].

Figure 2.4 Application of biodegradable polymers [8–15].

Other benefits that biodegradable polymers offer are increased soil fertility, low accumulation of plastic materials in the environment, and reduction in the cost of waste generated. However, some existing obstacles prove challenging for biodegradable polymers. These challenges are as follows:

1. The rate of biodegradation is often dictated by the type and ratio of the components in biopolymer.
2. Favorability of environmental conditions of biodegradation.

3. Inferior properties of biodegradable polymers in comparison with conventional plastics. Biopolymers are often blended with different additives to improve the properties.

Biodegradable polymers have found their use in applications such as packaging, medical, agriculture, and pharmaceutical. Some of the other applications where biodegradable polymers have been used are cosmetics, adhesives, lubricants, textiles, electronics, and high-strength structural materials [7]. In a biomedical industry, biodegradable polymers have found use in large implants, such as bone screws and bone plates, small implants, such as staples, sutures and drug delivery vehicles, plain membranes for guided tissue regeneration, and multifilament meshes or porous structures for tissue engineering as shown in Fig. 2.4 [8–15].

New biopolymers that can substitute or replace synthetically derived materials in traditional applications as well as range of new commercial opportunities are being investigated by chemical and biotechnology corporations [16].

2.2 Definitions

There has been a considerable shift from fossil-based resources to more environment-friendly resources because of their added benefits. In order to protect the environment, there has been a steady and rapid increase in the usage of products not only made from natural renewable resources but also products that decompose into environmental-friendly constituents. Globally, every country has set stringent regulations to restrict amount of solid waste generated by consumers. There are new programs that are initiated each year to offer consumer's benefits to make a switch from synthetic-based products to renewable resources.

In literature, there has been no consensus on the exact definition of the generic terms biodegradable, biopolymer, bio-based and oxy-degradable, which appear to have multiple and overlapping meanings. According to the European Union (EU) Norm EN 13432, titled "Requirement for packaging recoverable through composting and biodegradation: Test scheme and evaluation criteria for the final acceptance of packaging," has clearly stated standard set of criteria to determine whether a material can be considered compostable. An in-depth review of each term has been provided in the following sections.

2.2.1 Biodegradable

The terms biodegradation, biodegradable, and compostable are frequently misused and create misunderstanding. The European Norm EN 13432 has defined the term biodegradable as the one where degradation mechanism is characterized by the breakdown of organic chemical by microorganisms in the presence of oxygen to carbon dioxide, water, and mineral salts of any other element present (mineralization) and new biomass or in the absence of oxygen to carbon dioxide, methane, mineral salts, and new biomass. Microorganisms present in the disposal environment utilize the carbon product to extract chemical energy for their life processes. They do so by breaking the material into smaller molecules, transporting these small molecules inside the microorganisms cell and further oxidizing these small molecules to carbon dioxide and water, and releasing energy as shown in Fig. 2.5 [17–19].

The term biodegradable is also directly associated with different disposal ways such as composting, sewage treatment, denitrification, or anaerobic sludge treatment. The rate of degradation should be consistent with the disposal method. According to the European Norm EN 13432, compostable material is differentiated from biodegradable if their characteristics show [20]:

1. Biodegradability—In order to show complete biodegradability, a biodegradation level of at least 90% must be reached in less than 6 months.
2. Disintegrability, namely fragmentation and loss of visibility in the final compost.

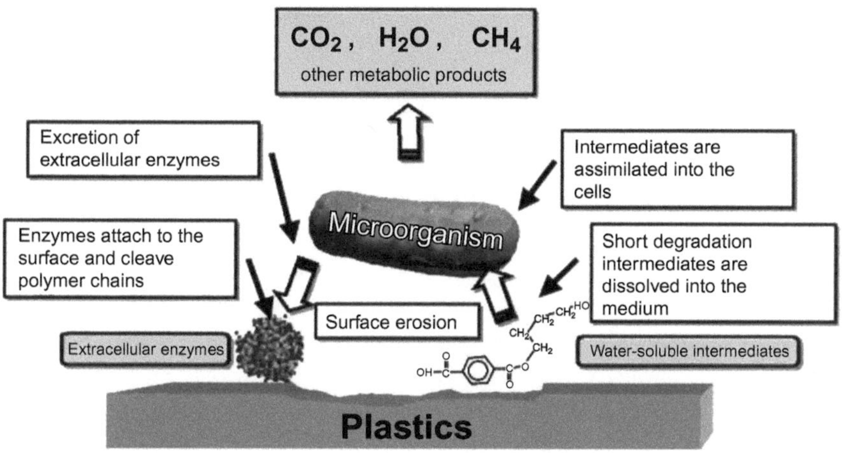

Figure 2.5 Biodegradation of polymeric materials [17].

3. Absence of negative effects on the composting process.
4. Low levels of heavy metals which are below given max values and the absence of negative effects on the final compost.

2.2.2 Biopolymer

Similar to any other polymers, biopolymers are chain-like molecules made up of repeating chemical blocks with very long lengths. The prefix "bio" indicates that they are produced by living organisms and are therefore biodegradable. Biopolymers can be classified in three groups, depending on the nature of the repeating unit they are made of: (i) polysaccharides are made of sugars, (ii) proteins of amino acids, and (iii) nucleic acids of nucleotides. Glycoprotein, which is a combination of protein and carbohydrate, plays a role in immune cell recognition and tissue adhesion. Nucleic acids are involved with the storage of the genetic code (DNA) and the translation of the genetic information into protein products (RNA) as shown in Fig. 2.6 [21].

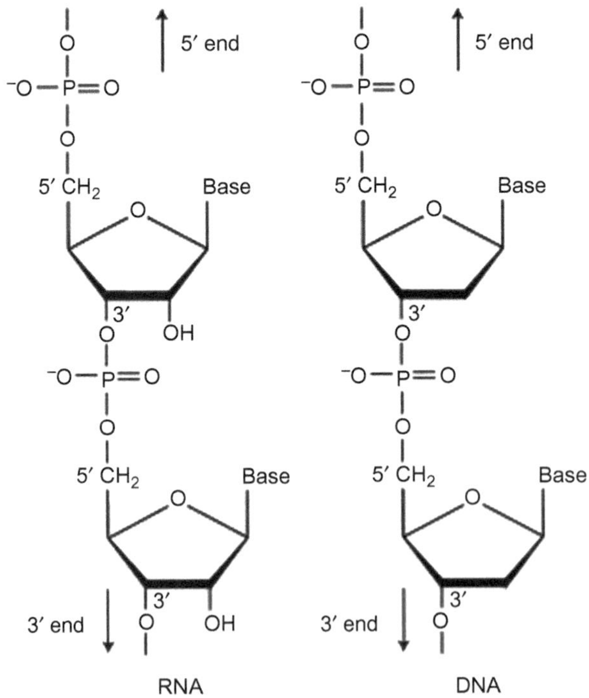

Figure 2.6 Chemical structures of ribonucleic acid (RNA) and deoxyribonucleic acid (DNA) [21].

Table 2.1 Biopolymers Found in Nature and Their Functions [22]

Polymer	Monomers	Function(s)
Nucleic acids (DNA and RNA)	Nucleotides	Carriers of genetic information universally recognized in all organisms
Proteins	Alpha-ammo acids	Biological catalysts (enzymes), growth factors, receptors, structural materials (wool, leather, silk, hair, connective tissue), hormones (insulin); toxins, antibodies
Polysaccharides (carbohydrates)	Sugars	Structural materials in plants and some higher organisms (cellulose, chitin); energy storage materials (starch, glycogen); molecular recognition (blood types)
Polyhydroxyalkanoates	Fatty acids	Microbial energy reserve materials
Polyphenols	Phenols	Structural materials in plants (lignin), soil structure
Polyphosphates	Phosphates	Inorganic energy storage materials
Polysulfates	Sulfates	Inorganic energy storage materials

Table 2.1 shows different kinds of naturally occurring biopolymers defined by the chemical structure of their monomeric units [22].

A typical biopolymer is made of blocks of monomer units connected to another in a linear chain; however in some cases, they can be branched or cross-linked. Biopolymers can also exist as heteropolymers because they may contain variety of monomeric units. Biopolymers have the ability to form hierarchical structures at successive lengths. Starting out at primary structures, the monomeric units are organized in a local molecular conformation. This local molecular conformation is referred to as secondary structure. Sometimes, biopolymer has the ability to take the shape of three-dimensional conformation known as tertiary structure.

Some of the advantages of biopolymers are as follows:

- Help in the reduction of carbon dioxide from atmosphere and carbon emissions. Carbon dioxide released during biodegradation process is reabsorbed by crops grown as a substitute in their place
- Compostable and help in reducing environmental pollution

- Biopolymers are renewable because of neutral carbon in them. Because they are composed of living organism, the process is completely sustainable
- Reduce dependency on fossil-based plastics

2.2.3 Bio-based

Definition of Bio-based plastics is drawn from the European Committee for Standardization (CEN) literature, which defines as the plastics that is derived from biomass [20,23]. In general, biomass is referred to biodegradable organic material derived from plants, animals, and microorganisms and is considered as renewable. Some plastics are fully bio-based and may be biodegradable, such as starch and polyhydroxyalkanoates; some may be partially bio-based and biodegradable such as polylactic acid and cellulose, whereas others may be partially bio-based and nonbiodegradable such as bio-polyethylenetetraphlate, bio-polypropylene, and bio-polyethylene. It is important to understand that the ability of the bio-based plastics to degrade does not depend on its bio-based content but rather on its structure and physical properties.

Some of the advantages of bio-based plastics are as follows:

- Bio-based plastics contain renewable carbon.
- Some types of bio-based plastics can be degraded through composting:
  - A benefit for applications such as disposal cutlery and flexible food packaging where the plastics can be disposed of alongside food wastes.
  - Biodegradable agricultural plastic mulches may also be simply ploughed into the field when needed, saving the need for collection and storage of plastics.
- Drop-in bioplastics such as bio-based polyethylene terephthalate (bio-PET) are similar to fossil-based plastics and can be recycled in the same way as conventional plastics.
  - At the end of their life, when recycling is no longer an option, these materials can be combusted to produce renewable energy and/or be used to make biofuels.
- Bio-based plastics can reduce energy consumption and carbon dioxide emissions compared to fossil-based plastics.

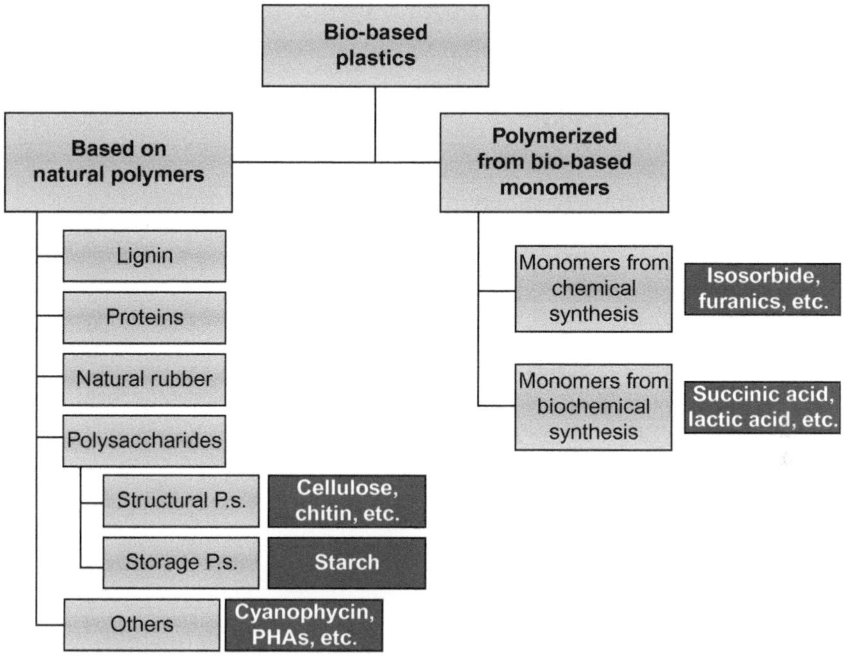

Figure 2.7 Overview of bio-based plastics [23].

Fig. 2.7 shows an overview of bio-based plastics [23], which can be distinguished based on natural polymers and the ones polymerized from bio-derived monomers.

2.2.4 Oxo-degradable

Oxo-degradable plastics are based on conventional plastics, like polyethylene (PE), polypropylene (PP), polystyrene (PS) and polyethylene terephthalate (PET), to which additives (also known as oxo-degradable additives) are added. This causes plastics to degrade by a process initiated by oxygen and accelerated by light and/or heat. These oxo-degradable additives are metal salts of carboxylic acids or dithiocarbamates based on cobalt (Co), iron (Fe), manganese (Mn), nickel (Ni), or Cerium (Ce), with Co being used more for packaging and Fe and Ni more for mulch film. The actual content of metal in a salt is typically less than 10% with the addition level of the active ingredients being approximately 0.1% of the finished film [24].

The general degradation mechanism for oxo-degradable plastics consists of three stages [25–30]:

1. The initiation stage

$$RH(heat, O_2, stress) \rightarrow ROOH$$
$$ROOH(heat\ and/or\ UV\ light) \rightarrow RO^{\bullet} + {}^{\bullet}OH$$
$$PH \rightarrow R^{\bullet} + POH(H_2O)$$

2. The propagation stage

$$R^{\bullet} + O_2 \rightarrow RO_2{}^{\bullet}$$
$$RO_2{}^{\bullet} + RH \rightarrow ROOH + R^{\bullet}$$

3. The termination stage

$$2R^{\bullet} \rightarrow R - R$$
$$R^{\bullet} + ROO^{\bullet} \rightarrow ROOR$$
$$2ROO^{\bullet} \rightarrow O_2 + ROH + R = O$$

The degradation process not only involves fragmentation but also changes the molecular structure of the material so that it ceases to be a plastic and becomes biodegradable. It then disappears completely without leaving any fragments or toxic residues, instead changing into carbon dioxide, water, and biomass.The end result is no plastic, no harmful residues, and no long-term damage at all. The length of useful life of an oxo-degradable plastic is determined by processing aids such as antioxidants, processing stabilizers, and UV stabilizers contained within the formulation and can be modified to match degradation timescale needed. Oxo-degradable plastics do not meet ASTMD6400 specifications for compostability and are therefore not approved by the Biodegradable Products Institute (BPI). According to the standard, the product is required to degrade within 180 days.

References

[1] M. Niaounakis, Biopolymers reuse, recycling, and disposal, Introduction to biopolymers, Plastics Design Library, William Andrew Publications, Boston, 2013 (Chapter 1).

[2] Polymer Structure, NDT Resource Center, <www.nde-ed.org/EducationResources/CommunityCollege/Materials/Structure/polymer.htm/>.

[3] L. Shen, J. Haufe, M.K. Patel, Retrieved from: <www.epnoe.eu> Product Overview and Market Projection of Emerging Bio-Based Plastics, Utrecht University, Utrecht, The Netherlands, 2009.

[4] L. Averous, Biodegradable multiphase systems based on plasticized starch: a review, J. Macromol. Sci. Polym. Rev. Part C 44 (3) (2004) 231–274, ISSN 1532–1797.

[5] A. Ashwin Kumar, K. Karthick, K.P. Arumugam, Biodegradable polymers and its applications, Int. J. Biosci. Biochem. Bioinf. 1 (3) (2011) 173–176.

[6] Environment, <https://www.ied.edu.hk/biotech/eng/classrm/class_env2.html>.

[7] R. Chandra, R. Rustogi, Biodegradable polymers, Prog. Polym. Sci. 23 (1998) 1273–1335.

[8] B. Ghanbarzadeh, H. Almasi, Biodegradable Polymers, InTech Publishers, 2013, pp. 141–185 (Chapter 6).

[9] L. Averous, E. Pollet, Biodegradable Polymers, Environmental Silicate Nano-Biocomposites, Green Energy and Transport, Springer-Verlag London, 2012.

[10] RESOMER® Biodegradable Polymers, Evonik Industries AG—Pharma Polymers and Services. <http://www.pharmaceutical-technology.com/contractors/excipients/evonik-pharma-polymers/evonik-pharma-polymers5.html/>.

[11] M. Matthew, B.A. Robbins, A.R. Vaccaro, L. Madigan, The use of bioabsorbable implants in spine surgery, Neurosurg. Focus 16 (2004) 3.

[12] Business Insider, <http://www.businessinsider.com/scientists-create-biodegradable-computer-chip-out-of-wood-2015-5>.

[13] Integral BioSystems, Microencapsulation and Nano-encapsulation. <http://www.integralbiosystems.com/drugdelivery.php>.

[14] Biodegradable Sneakers, European Bioplastics, <http://en.european-bioplastics.org/press/press-pictures/textilesfibers/>.

[15] Making Plastic Bottles Environmentally-Friendly, <http://ensoplastics.com/ENSO/biodegradable-water-bottles.html>.

[16] Biopolymers: Making Materials Nature's Way, US Congress, Office of Technology Assessment, OTA-BP-E-102, US Government Printing Office, Washington DC, 1993.

[17] R.J. Muller, Biodegradability of polymers: regulations and methods for testing, in: A. Steinbuchel (Ed.), Biopolymers—General Aspects and Special Applications, vol. 10 (19), Wiley-VCH Verlag GmbH & Co. kGaA, Germany, 2005.

[18] R. Narayan, Biodegradability, Bioplastics Magazine 4 (2009) 28–31.

[19] R.-J. Muller, Biodegradability of polymers: regulations and methods of testing, Biopolymer Online (2005) 365–374.

[20] F. Degli Innocenti, Original press release, Environmental affairs, Novamont, 2004.

[21] R.C.J. Van der Maarel, Biopolymers, Introduction to Biopolymer Physics, World Scientific Publishing, 2008 (Chapter 1).
[22] U.S. Congress, Office of Technology Assessment, Introduction, Biopolymers: Making Materials Nature's Way, OTA-BP-E-102, U.S. Government Printing Office, Washington, D.C, 1993 (Chapter 1).
[23] S. Kabasci, Bio-based Plastics: Materials and Applications, Wiley & Sons, 2014.
[24] S. Deconinck, B. De Wilde, Benefits and challenges of bio- and oxo-degradable plastics, 2013, 50.
[25] G. Scott, D. Gilead, Degradable Polymers: Principles and Applications, Chapman & Hall, London, 1995.
[26] G. Scott, Degradation and stabilization of carbon-carbon chain polymers, Degradable Polymers: Principles and Applications, Kluwer Academic Publishers, The Netherlands, 2002 (Chapter 3).
[27] G. Scott, D.M. Wiles, Programmed-life plastics from polyolefins: a new look at sustainability, Biomacromolecules 2 (2001) 615–622.
[28] G. Scott, Abiotic control of polymer biodegradation, Trends Polym. Sci. 5 (1997) 361–368.
[29] G. Scott, D. Gilead, Introduction to the abiotic degradation of carbon chain polymers, Degradable Polymers: Principles and Applications, Chapman & Hall, London, 1995 (Chapter 1).
[30] D.M. Wiles, Oxo-biodegradable polyolefins in packaging, Biodegradable Polymers for Industrial Applications, Woodhead Publishing Limited, London, 2005 (Chapter 16).

3 Mechanisms of Polymer Degradation

3.1 Introduction

Polymers are a broad class of materials that are made from repeating units of smaller molecules called monomers. Polymers can be natural in origin or synthetic. Polymer obtained from lignin of the tree branches is considered natural while anything that is made by humans from naturally occurring materials is considered synthetic [1]. Very small variations in the chemical structures of polymer could lead to large changes in their biodegradability. Biodegradability depends on the molecular weight, molecular form, structure, and crystallinity. It decreases with increase in molecular weight, while monomers, dimers, and repeating units degrade easily.

There are three main biodegradation mechanisms [2] that are used to degrade biopolymers. They are:

- Microbial biodegradation
- Aerobic biodegradation
- Anaerobic biodegradation

Microbial degradation takes place through the action of enzymes or by-products generated by microorganisms such as bacteria, yeasts, and fungi. Two key steps happen in the microbial polymer degradation process: (1) depolymerization and (2) mineralization.

Fig. 3.1 shows reaction pathways during biodegradation of polymers [3]. Depolymerization is a free-radical mechanism in which the polymer is degraded to monomers, oligomers, and dimers which when exposed to an aerobic environment will yield carbon dioxide and water [3]. However, when the same depolymerized oligomers, dimers, and monomers are exposed to anaerobic environment, the end products are methane, carbon dioxide, and water [3]. Depolymerization occurs outside the organism due to the size of the polymer chain. Extracellular enzymes are responsible for this step [4].

Synthetic polymers such as poly(caprolactone) (PCL) are depolymerized by microbial enzymes abiotically, after which biodegradation of

Introduction to Bioplastics Engineering.
DOI: http://dx.doi.org/10.1016/B978-0-323-39396-6.00003-8

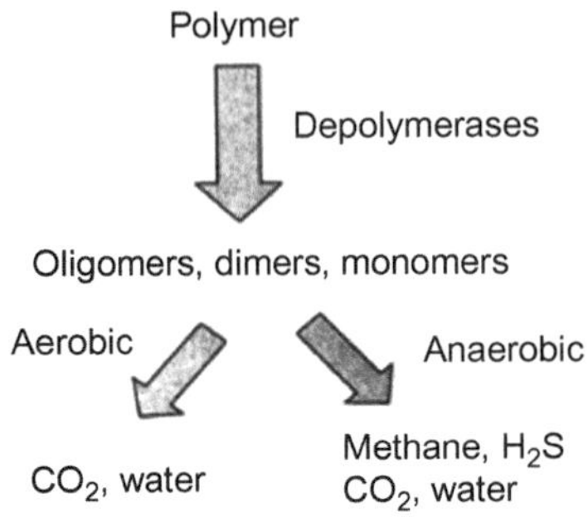

Figure 3.1 Reaction pathways during biodegradation of polymers [3].

the monomers by microorganisms proceeds as a secondary process. Abiotic hydrolysis is the most important reaction for initiating the environmental degradation of synthetic polymers such as polycarboxylates, poly(ethylene terephthalate) (PET), polylactic acid (PLA), and their copolymers. For all of these materials, hydrolysis acts as the initial step of splitting the polymer into its monomers, after which the monomers can be biodegraded [5]. Abiotic oxidation can also initiate the degradation of some polymers. As an example, polyethylene undergoes an auto-oxidation, which gradually reduces its molecular weight to the point where biodegradation can proceed [6].

Mineralization is the process of conversion of organic products to inorganic. Thus, in the mineralization or organic C, N, P, S, or other elements, carbon dioxide (CO_2) or inorganic forms of N, P, S, or other elements are released by the organisms and enter the environment. However, in mineralization of synthetic chemical, microorganisms are frequently the sole means of converting synthetic chemical to inorganic product. Mineralization results in the total destruction of the parent product [7].

Aerobic biodegradation is the process of breaking down organic contaminants by microorganisms during the presence of oxygen. Organic contaminants are rapidly degraded under aerobic conditions by aerobic bacteria called *aerobes*. Anaerobic degradation generally occurs when anaerobic microbes are dominant over aerobic microbes such as *Clostridia*, *Eubacterium* spp. Typically, there are four key stages of anaerobic degradation [8]:

- Hydrolysis
- Acidogenesis
- Acetogenesis
- Methanogenesis

Table 3.1 Modes and Corresponding Factors of Degradation [9,10]

Mode of degradation	Factors
Thermal degradation	Exposure to heat
Thermo-oxidative degradation	Exposure to heat and oxygen
Photo-degradation	Exposure to visible light and ultraviolet (UV) light
Irradiation degradation	Exposure of high-energy radiation such as X-rays and gamma irradiation
Mechanochemical degradation	Exposure to mechanical stress
Chemical degradation	Exposure to chemical attack such as solvolysis/ hydrolysis, ozonolysis, catalytic degradation
Biodegradation	Exposure to aerobic and anaerobic environment

Table 3.1 provides different modes of degradation and their contributing factors [9,10].

3.2 Polyesters

Polyesters are polymers in which component monomers are bonded via ester linkages. These ester linkages are generally easy to hydrolyze and hence a number of synthetic polyesters are biodegradable. Therefore, bacterial polyesters (polyhydroxyalkanoates, PHAs) have been used to make biodegradable plastics. Hydrolytic cleavage of the ester bond in the low molecular weight polyester by the lipase of *Pseudomonas* sp. is well known. The generalized structure of polyester is shown in Fig. 3.2 [11].

In general, polyesters are hygroscopic in nature and thus sensitive to moisture. PET is a very common polymer of polyester family. When an amorphous PET is exposed to high temperature, it undergoes hydrolytic degradation. A generalized reaction representing polycondensation and hydrolysis of polyesters is shown in Fig. 3.3 [12].

In the figure, R and R′ represent aromatic and aliphatic groups, respectively. Both polycondensation and hydrolysis happen above the melting point of polyester. The bulk material is only attacked at the surface when the temperatures are below melting point of the polymer and there are no solvents involved.

Fig. 3.4 shows ester linkage which binds together two monomer components together to form polyester [13]. When ester linkages get

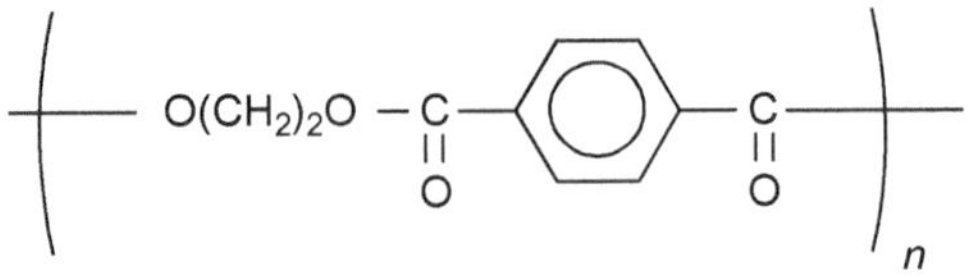

Figure 3.2 Generalized structure of polyester [11].

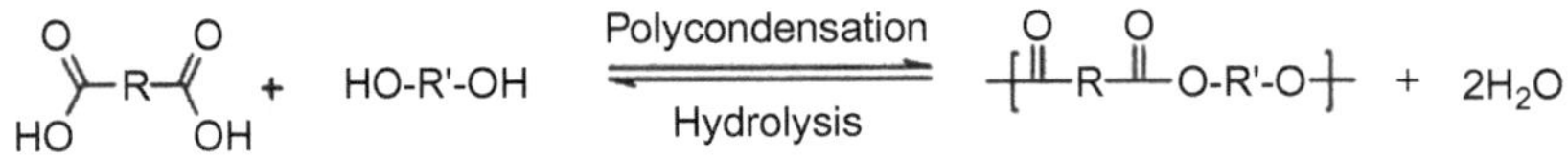

Figure 3.3 A generalized reaction showing polycondensation and hydrolysis of polyesters [12].

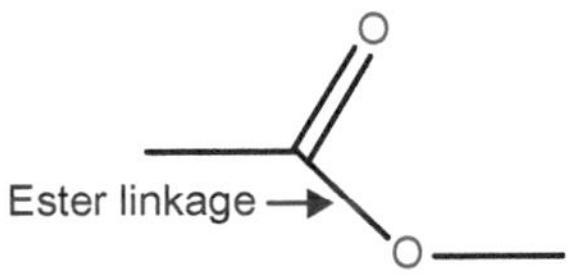

Figure 3.4 Ester linkage [13].

exposed to an enzyme, they are broken down. These enzymes termed as esterase splits esters into an acid and an alcohol in a chemical reaction with water in a process known as hydrolysis as shown in Fig. 3.3. Using this process, number of synthetic polyesters can be hydrolyzed to make them biodegradable. As an example, PHAs have been successfully made into biodegradable plastics using this approach. In other cases, lipases from *Pseudomonas* have been used to hydrolyze low molecular weight polyesters [13].

For semicrystalline polyesters in aqueous solution, degradation happens in two stages: (1) diffusion of water and (2) degradation. It all starts when water diffuses into randomly oriented amorphous regions of the polyester. Water is able to penetrate more easily because amorphous regions are less organized. The process of degradation is initiated at the second stage where amorphous regions are attacked by water undergoing hydrolysis. The hydrolysis starts at the edge and then proceeds to the center of the crystalline domains. The rate of this hydrolytic attack is much faster for amorphous polyesters as compared to semicrystalline polyesters [14].

There are two different mechanisms that are associated with biodegradation of polyesters: (1) bulk degradation and (2) surface erosion [15]. Bulk degradation process is initiated when water diffuses into the polymer matrix faster than the polymer is degraded. This results in the

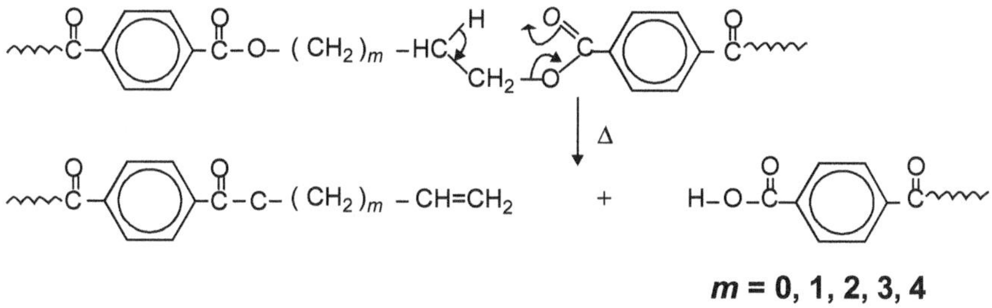

Figure 3.5 General degradation mechanism of aromatic polyesters [18–21].

hydrolyzable bonds in the polymer matrix to get cleaved homogeneously resulting in decrease in the average molecular weight of the polymer. The process of degradation through surface erosion happens when the diffusion rate of water into the polymer matrix is slower than the rate of degradation of the polymer. The degradation rate is dependent on the size of the exposed surface, resulting in a heterogeneous process. The degradation only happens on the thin surface layer resulting in average molecular weight of the polymer to remain unchanged [16]. Bulk degradation process happens in majority of polyesters with an exception of polyanhydride materials which differ from the common polyesters by the fact that they undergo degradation through surface erosion mechanisms [17].

Fig. 3.5 shows general degradation mechanism of aromatic polyesters [18–21]. Thermal degradation of aromatic polyesters at high temperature has been extensively studied. In addition to the list of aromatic polyesters, PET and poly(butylene terephthalate) have also been studied. As shown in Fig. 3.5, the C—O bonds located at position from C=O bonds are subjected to thermal cleavage [18–21].

Fig. 3.6 shows a two-stage thermal degradation mechanism of aromatic polyester [22–25]. Poly(diphenolic acid-isophthaloyl chloride) is exposed to elevated temperatures of higher than 400°C. The first stage involves six steps. Dehydration reaction between carboxyl groups happens during the first step generating anhydride groups and water. The second step involves elimination of carbon dioxide from the anhydride groups forming ketone bonds [22–25]. Further elimination of carbon monoxide leads to the formation of double C=C bonds in the third step of the first stage. Next step of this stage is the elimination of ethylene as shown in reaction (d) of Fig. 3.6. Other reactions in this stage include dehydration reaction between phenolic hydroxyl groups and transesterification between anhydride and hydroxyl groups from which water is produced and ether and ester bonds are formed, respectively.

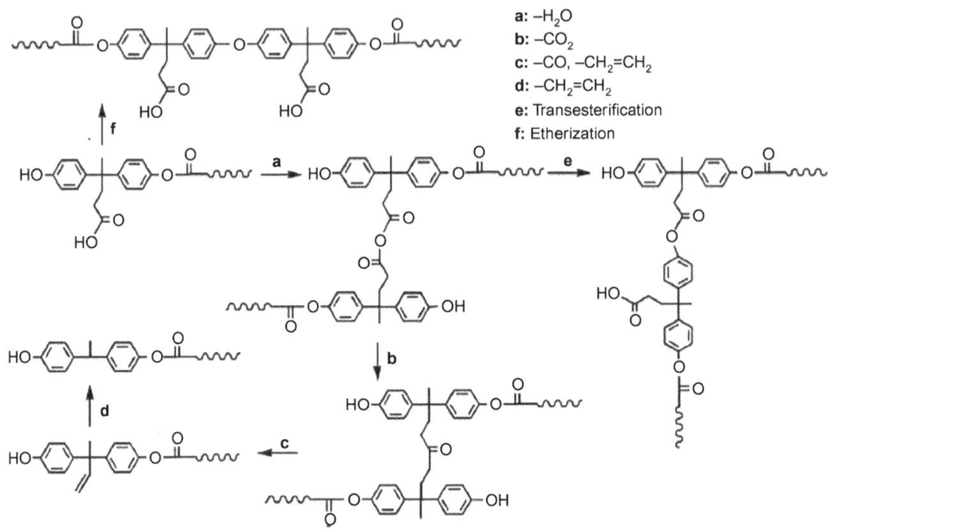

Figure 3.6 Two-stage thermal degradation mechanism of polyester [22–25].

In second stage, ordinary thermal oxidization degradation of the polymer backbone takes place and large amount of carbon dioxide is released [26].

In case of aliphatic polyesters, degradation proceeds by one or more mechanisms which include hydrolysis, microbial, enzymatic, and thermal. The degradation happens either at the surface or within the bulk [25]. In a typical aliphatic polyester such as poly(propylene succinate) (PPSu), the degradation happens by two consecutive mechanisms [26]. Similar to other aliphatic polyesters such as PCL, PPSu sees decrease in its molecular weight, while water, carbon dioxide, and 5-hexamoic acid are released as a by-product [27].

Fig. 3.7a–d presents different stages of the degradation mechanism of aliphatic polyester [26]. Fig. 3.7a represents the first stage of the degradation mechanism. During the first stage, the gases that are produced from the decomposition of hydroxyl and carboxylic end-groups of polyesters cause increase in the degradation temperature. Among the released gases, carbon dioxide is slightly reduced while hydrogen increases with increase in the degradation temperature. As the temperatures exceed 450°C, it will result in homolytic cleavage of ester groups and produce large amounts of carbon monoxide and methane.

Fig. 3.7b is the second stage of the degradation mechanism. It is worthy to note that the by-products are formed by the β-hydrogen bond scission of the propanediol monomer. At low degradation temperatures,

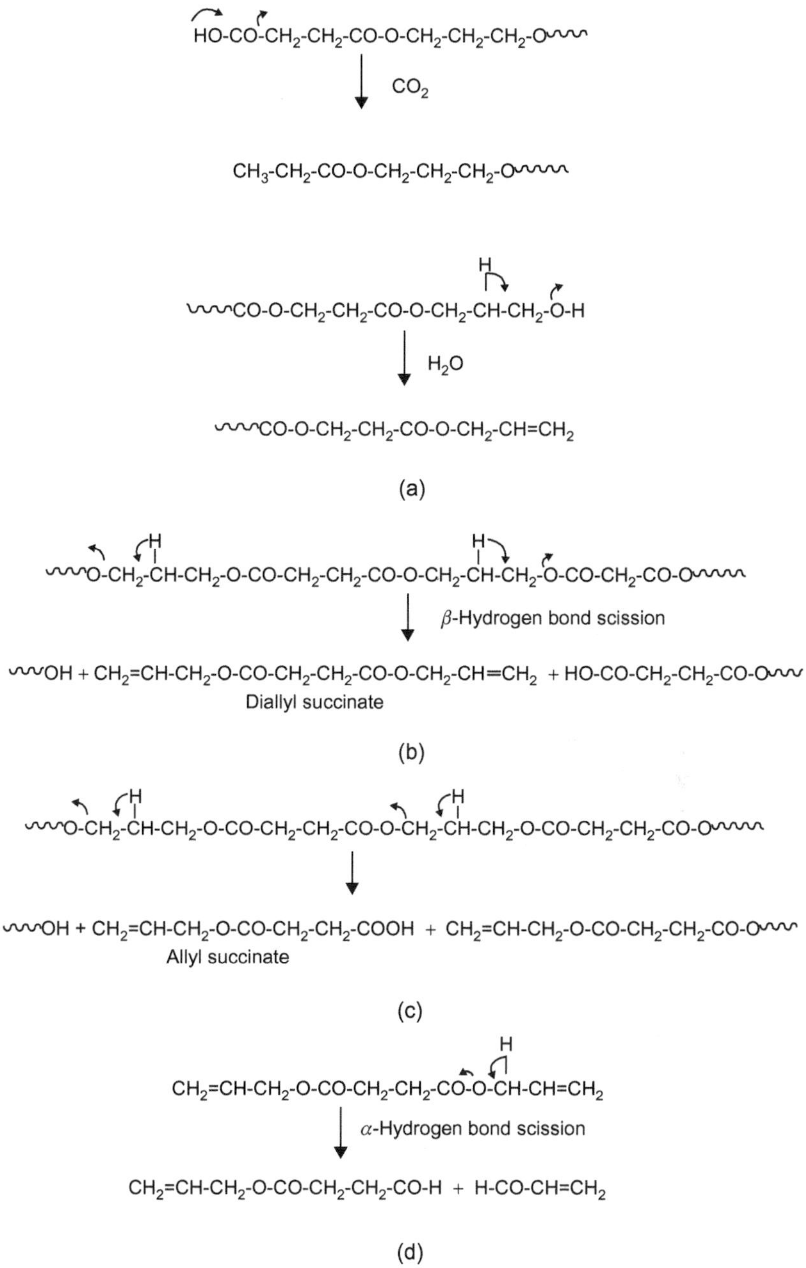

Figure 3.7 General degradation mechanism of aliphatic polyesters [26].

the bond scission of chains happens simultaneously at positions just right and left from a succinic acid monomer, producing diallyl succinate. However, allyl succinate is produced if one of the two chain scissions does not follow same reaction path as shown in Fig. 3.7c. Fig. 3.7d shows a reaction in which diallyl succinate is further degraded at low decomposition temperatures producing two aldehydes. These aldehydes are produced not by β-hydrogen bond scission but by a hydrogen bond scission, as shown in Fig. 3.7d [26].

3.3 Polyamides

Polyamides are a group of polymers which contain an amide group in their repeating unit. They can be produced by either ring-opening polymerization of cyclic lactames (eg, polyamide 6 (PA-6) and polyamide 11 (PA-11) or from condensation reactions between diacarboxylic acids and diamines (eg, polyamide 6,6 (PA-6,6) and polyamide 4,6 (PA-4,6)). Structures of some of the commonly available nylons are shown in Fig. 3.8 [27].

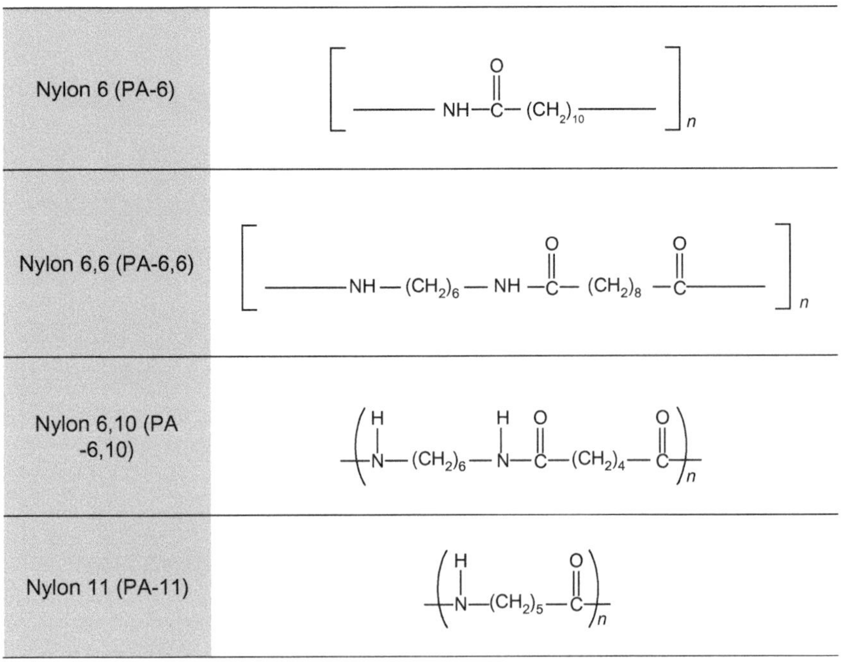

Figure 3.8 Different types of nylon [27].

All polyamides are hygroscopic and will absorb moisture when left open to the atmosphere. The degradation of polyamides is a complex process which can lead to many different products. There are different way polyamides degrade, for example, mechanical degradation from wear, biological degradation caused by micro- and macroorganisms, and chemical degradation caused by UV light, heat, or hydrolysis. Some of the main types of degradation processes for polyamides which will be discussed in this section are:

1. Thermal degradation
2. Photo-oxidation
3. Thermal oxidation

3.3.1 Thermal Degradation

All polyamides are hygroscopic and are susceptible to thermal degradation. Fig. 3.9a and b shows proposed monomer formation mechanisms

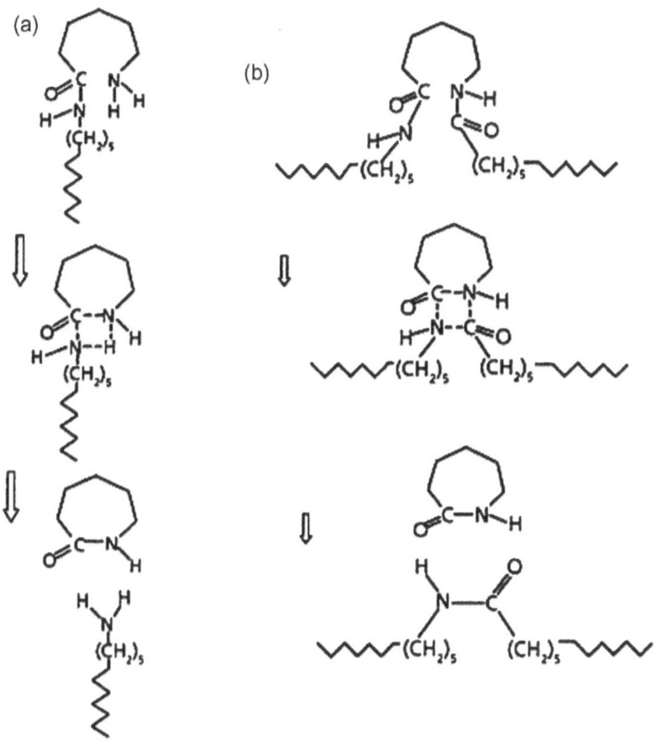

Figure 3.9 Proposed monomer formation mechanisms in PA-6: (a) initial fast rate and (b)*** slow rate [28].

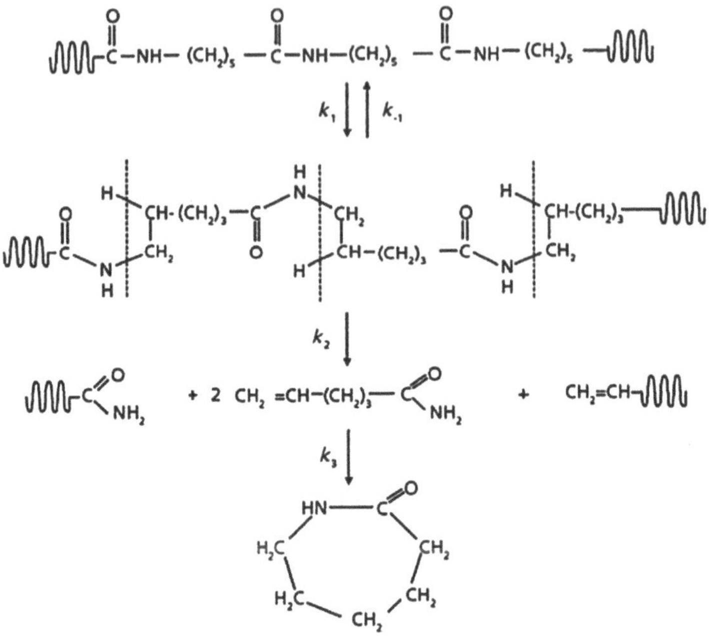

Figure 3.10 Degradation of PA-6 via cis-elimination [28].

in PA-6 at fast and slow rate of thermal degradation. A cyclic monomer also known as caprolactam is the main product formed as a result of degradation. However, the formation of this cyclic monomer will depend on the sample size and the experimental conditions otherwise will yield a different strain of product [28]. Besides formation of caprolactam, the depolymerization also yields oligomers with nitrile and vinyl chain ends. Temperatures above 420°C result in degradation of PA-6 to a higher order.

Fig. 3.10 shows degradation mechanism of PA-6 using cis-elimination reaction. In an elimination reaction, the atoms that are lost come from adjacent carbon atoms. The elimination reactions involve three steps:

1. Removal of proton
2. Formation of the C—C bond
3. Breaking of the bond to the leaving group

The cis-elimination reaction proceeds with a six-membered intermediate product shown in Fig. 3.10, consistent with kinetic data of ε-caprolactam provided in Table 3.2 [28].

Table 3.2 Kinetic Data on the Thermal Degradation of Polyamides in Nitrogen [28]

Parameter	PA-6	PA-6,6	PA-12	PA-6,12
E_{a1} (kJ/mol)	162	91	2208	164
$Log_{10}(k_0)$ 1	9.42	2.62	160	8.67
Order 1	1.34	4.57	8.10	1.97
$Log_{10}(k_{cat})$ 1	0.35	2.03	−19.68	1.35
E_{a2} (kJ/mol)	476	310	260	400
$Log_{10}(k_0)$ 2	32.95	21.08	16.47	27.25
Order 2	1.35	1.44	0.63	1.08

Different types of polyamides produce different products as a result of thermal degradation. For example, PA-6,10 produces mostly linear or cyclic oligomeric monomer when they are exposed to heat. There are five main possible mechanisms leading to five different product distributions. These mechanisms are:

1. Primary polyamide bond scission (C(O)—NH or NH—CH_2)
2. Hydrolysis
3. Hemolytic scission
4. Intramolecular C—H transfer
5. Cis-elimination

For PA-12, thermal degradation results in lactams as primary product with traces of olefinic nitriles and α-olefins. Thermal degradation of PA-6,12 results in the formation of cyclic oligomers with some traces of dinitriles.

3.3.2 Photo-Oxidation

Photo-degradation of a polymer typically happens by either chain scission or cross-linking reaction which occurs by the activation of the polymer molecule provided by absorption of light by the polymer. Proton-initiated degradation process starts when the light is absorbed by photo-initiators followed by photo-cleavage of polymer molecule into free radicals further initiating degradation of the polymer molecules [29]. The degree of breakage of polymer molecule depends on the wavelength of the UV light and the atmospheric conditions. Oxygen is necessary at longer wavelengths in order for photo-degradation to occur, whereas at shorter wavelength it not necessary. This conforms to the concepts of photosensitized oxidation and photolysis [30–31].

Process of photolysis happens at wavelengths below 3000 A resulting in direct scission of the polyamide chains and cross-linking. Specifically, the cross-linking is clearly visible in the formation of the phenolin-soluble material produced on exposure of nylon films to the unfiltered beam from the UV lamp.

The probable scissions of the nylon 6,6 chains are shown in Fig. 3.11 [32].

As shown in Fig. 3.11, equation (21) shows that the reaction produces acetic acid on hydrolysis; however, acetyl groups may have already been present in the un-degraded polymer due to the added acetic acid as stabilizer during its manufacture. This reaction is accompanied by unsaturated fragment insofar as traces of a C. An equation (22) through (24) explain the production of *n*-pentanoic acid on hydrolysis of photo-degraded nylon 6,6. With nylon 6,10 and nylon 6, the corresponding products would be *n*-nonanoic acid and *n*-pentylamine.

Fig. 3.12 shows the decomposition of the derived alkoxy radicals. It is observed from equations (25) and (26) that the oxidative attack

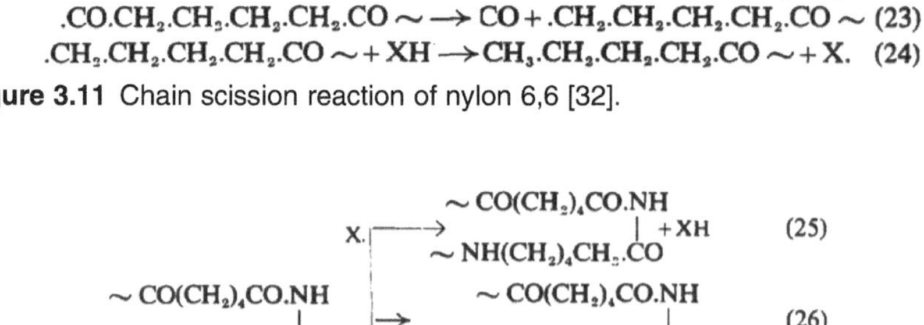

Figure 3.11 Chain scission reaction of nylon 6,6 [32].

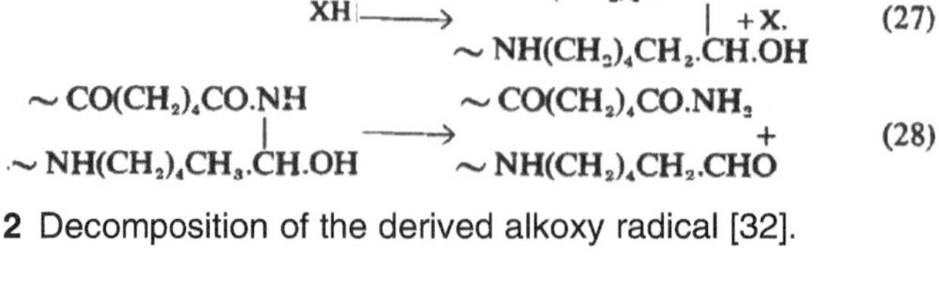

Figure 3.12 Decomposition of the derived alkoxy radical [32].

occurring predominantly at the methylene group adjacent to the N-atom with a subsequent decomposition of the derived alkoxy radical which follows thereafter [32].

Equations (27) and (28) show production of aldehyde end-groups in the polymer and ammonia on hydrolysis of the corresponding amide end-groups formed. Equation (26) also yields ammonia on hydrolysis together with formic acid and *n*-pentylamine [32].

Although these reactions dominate the photo-degradation of polyamides, they are not the only ones. In one of the other reactions, the hydrolysis of degraded nylon 6,6 results in the formation of small amounts of carbon monoxide and carbon dioxide from the decomposition of hydroxy- or keto-acids, in addition small amounts of glutaric and succinic acids [32,33].

3.3.3 Thermal Oxidation

Thermal oxidation of polyamides is primarily divided into three stages [34]:

Stage 1: Formation of *N*-acylamides

$$R-CO-NH-CH_2-R' \longrightarrow R-CO-NH-CO-R'$$

Stage 2: Formation of *N*-formamides as a result of C1−C2 scission

$$R-CO-NH-CH_2-R' \longrightarrow R-CO-NH-CHO$$

Stage 3: Oxidative dealkylation to yield carbonyl compound

$$R-CO-NH-CH_2-R' \longrightarrow R-CO-NH_2 + R'-CHO$$

Fig. 3.13 shows a revised mechanism for the thermal oxidation of aliphatic polyamides formulated by Karstens and Rossbach [34,35]. Karstens and Rossbach followed similar mechanism which was based on attacking on the *N*-vicinal methylene group. In this mechanism, the radical is formed during the initial step at the *N*-vicinal methylene group and then combines with oxygen to give a new radical. This new radical isomerize follows various reaction pathways, each involving chain scission resulting in the formation of carbonyl and carboxy end-groups [34,35].

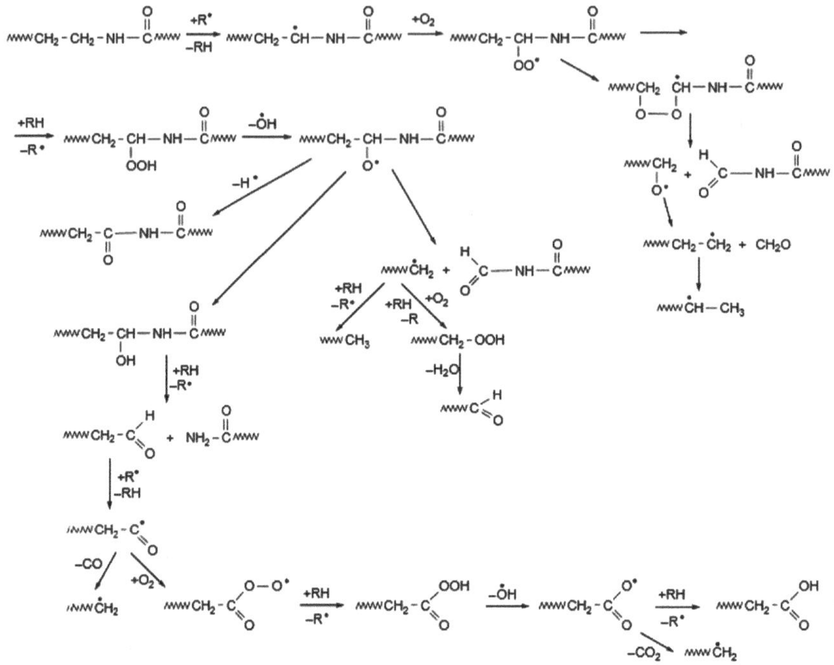

Figure 3.13 Mechanism for the thermal oxidation of polyamides [34,35].

3.4 Polycarbonates

This group of plastics is classified as engineering thermoplastics because of their high-performance characteristics. Polycarbonates are tough, rigid, and dimensionally stable. As with most other plastics containing aromatic groups, radiation stability is high [36,37]. The generalized chemical structure is shown in Fig. 3.14.

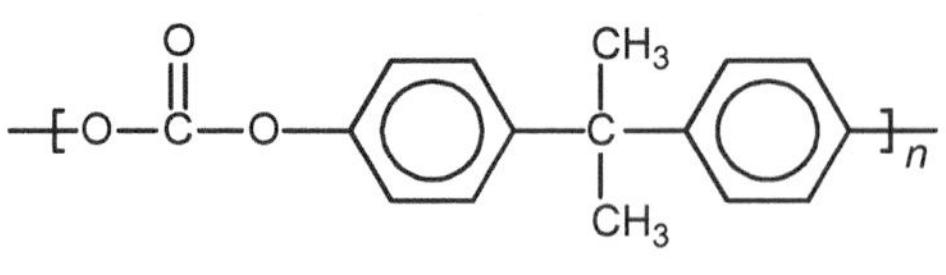

Figure 3.14 A generalized chemical structure [12,37].

Davis and Golden [37–39] were the first researchers to work on the chemical mechanism of polycarbonate degradation. Their work relied on techniques such as intrinsic viscosity, mass spectrometry, and gas chromatography.

They found out that polycarbonate undergoes very little decomposition below 250°C as shown in Fig. 3.15. They observed that if the

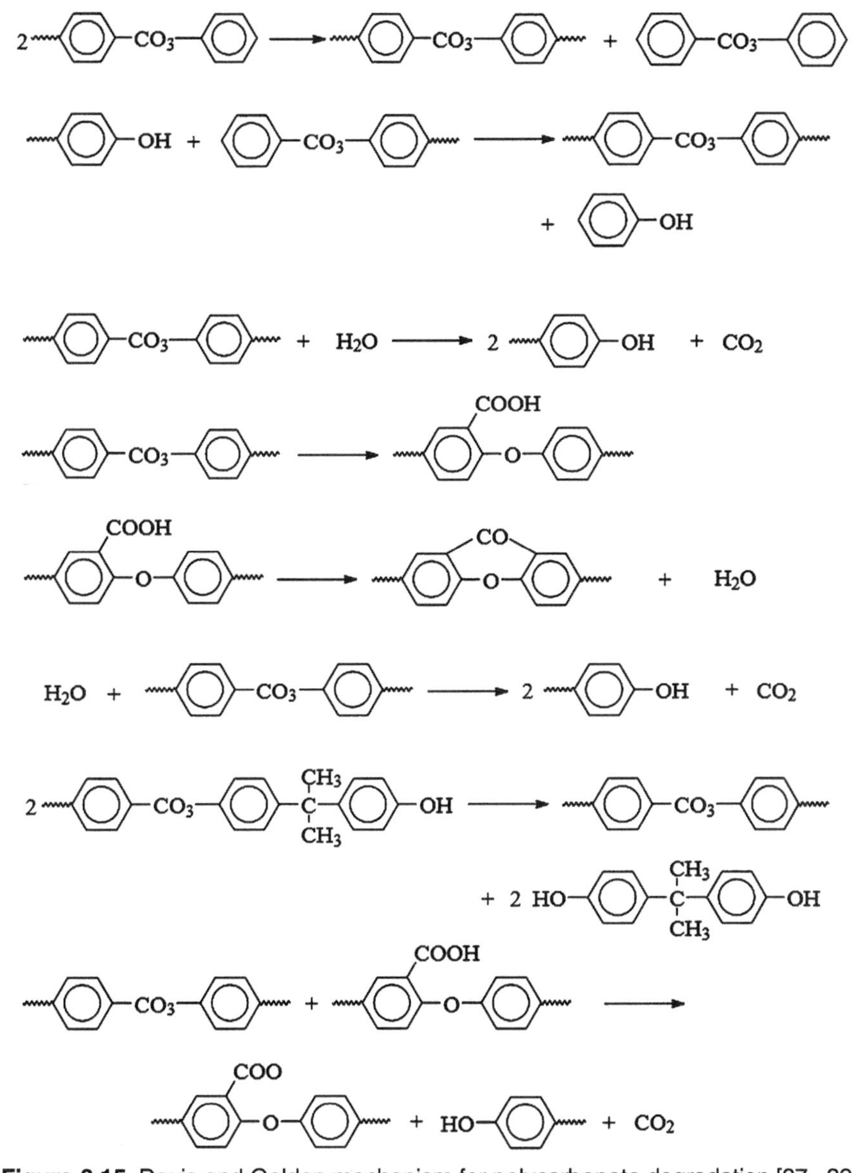

Figure 3.15 Davis and Golden mechanism for polycarbonate degradation [37–39].

system is continuously evacuated by removing volatile products, the material undergoes branching and eventually gets cross-linked resulting in insoluble gel formation. If these volatile products are not removed, this leads to chain scission. These two modes of degradation are the result of competition between condensation and hydrolysis.

Bisphenol-A and carbon dioxide constitute two main volatile products of polycarbonate. Lesser amounts of carbon monoxide, methane, diphenyl carbonate, phenol, and 2-(4-hydroxyphenyl)-2-phenyl propane are also produced. When bisphenol-A is further broken down, it produces ethyl phenol, isopropenylphenol, isopropylphenol, and cresol.

It is reported that the carbonate group is the major point of degradation at temperatures below 400°C, while at higher temperatures, the isopropylidene group is also susceptible to loss of a methyl radical [37–39].

Fig. 3.16 shows degradation mechanism proposed by Lee [40]. He proposed that when the temperature is in the range of 340–380°C, polycarbonate gets depolymerized. This depolymerization consists of hydrolysis and alcoholysis of the carbonic ester. Carbonic ester is a form of ester exchange where chains break near the middle. The hydrolysis produces carbon monoxide and either bisphenol-A (if it occurs near the chain end) or two shorter chains. Complex random chain scission dominates at higher temperatures in the range of 480–600°C. Under certain conditions, random chain scission can include decarboxylation, hydrolysis, hydrogen abstraction, chain scission and ether cleavage [40].

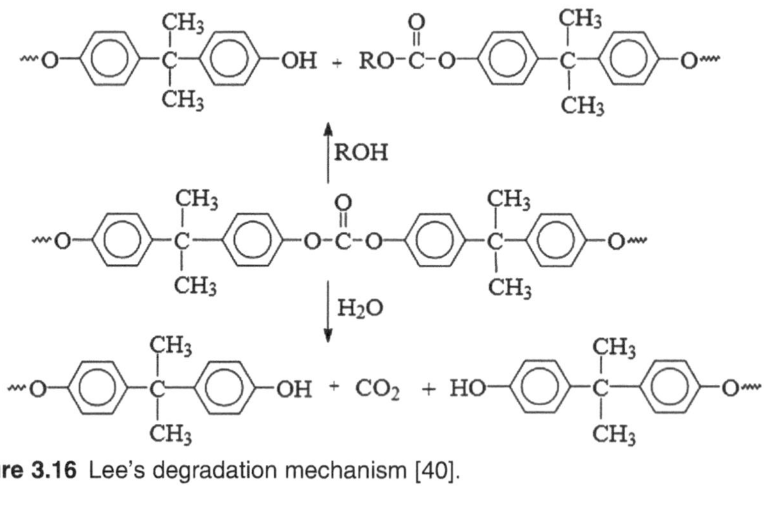

Figure 3.16 Lee's degradation mechanism [40].

Fig. 3.17 shows Kuroda's mechanism for polycarbonate degradation [41]. This mechanism presents a radical reaction scheme for degradation at the isopropylidene group as well as scheme for cross-linking. Kuroda hypothesized that the carbonate linkage and isopropylidene group in polycarbonate are most susceptible to cleavage while ester linkage is assumed to be thermally stable and point of any cross-linking [41].

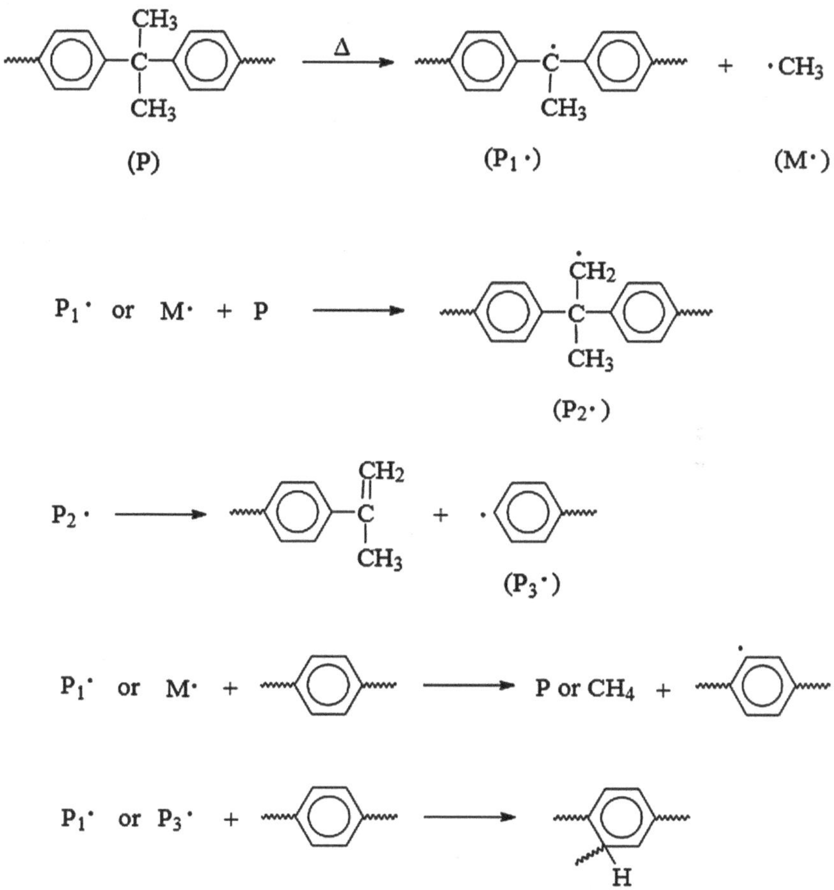

Figure 3.17 Kuroda's mechanism for polycarbonate degradation [41].

Part figures a–d of Fig. 3.18 are different mechanisms of polycarbonate degradation proposed by McNeill and Rincon [42,43]. As shown in Fig. 3.18a, phenol and cyclic dimer are early degradation products. The cyclic dimer is produced by the radical depolymerization that involves scission at isopropylidene. The second proposed mechanism shown in Fig. 3.18b illustrates homolytic scission at the carbonate linkage producing bisphenol-A. Fig. 3.18c shows the third proposed mechanism for polycarbonate degradation. In this mechanism, the migration of methyl groups from homolytic reactions results in the production of *p*-cresol and *p*-ethyl phenol. The fourth mechanism shown in Fig. 3.18d illustrates hydrogen abstraction after hemolysis in the methyl group. This reaction leads to cross-linking, scission, and methyl group migration producing *p*-vinyl phenol and *p*-cresol.

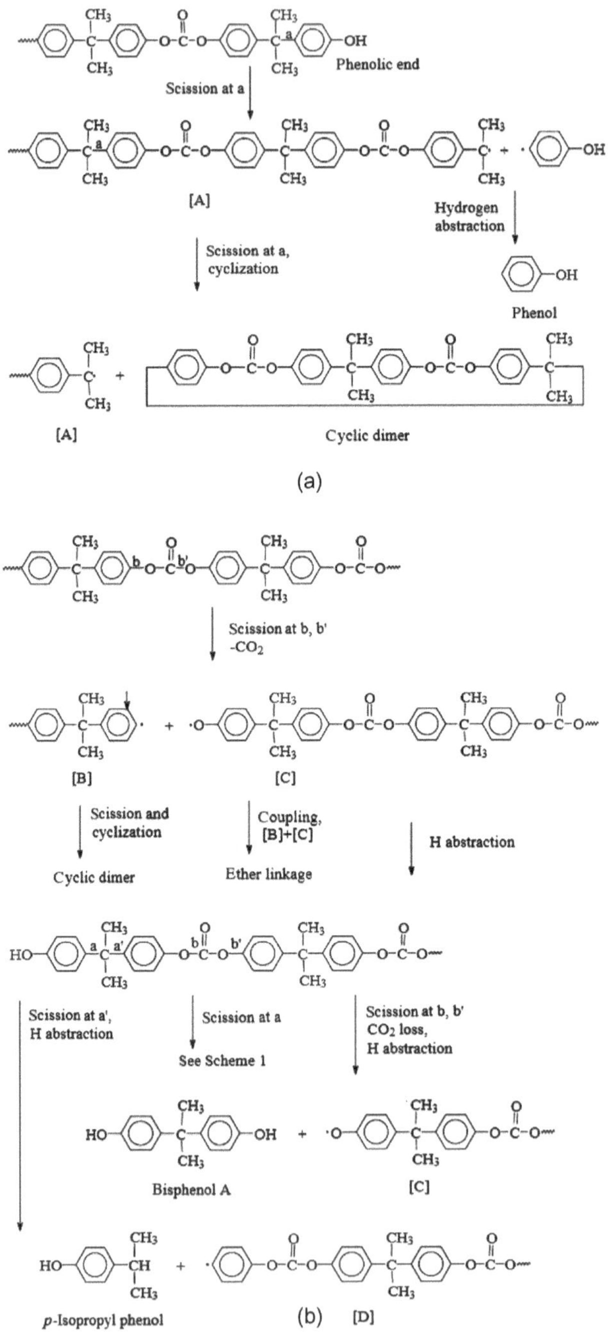

Figure 3.18 (Continued)

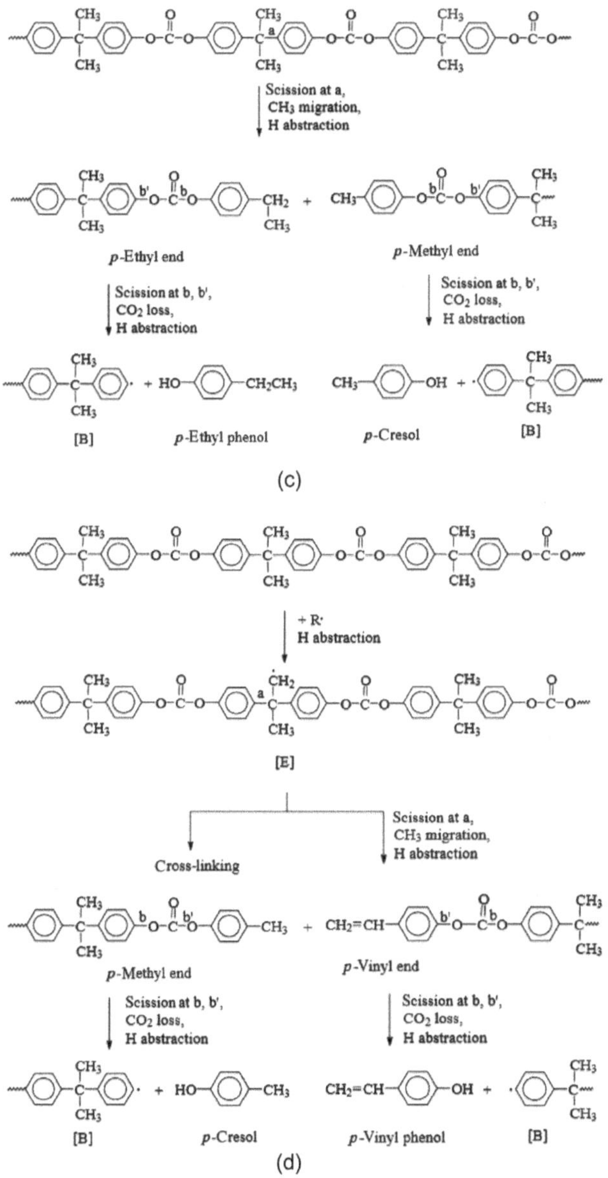

Figure 3.18 McNeill and Rincon's (a) mechanism 1, (b) mechanism 2, (c) mechanism 3, and (d) mechanism 4 for polycarbonate degradation [42,43].

Fig. 3.19a shows a mechanism for polycarbonate degradation as proposed by Montaudo and Puglisi. In this mechanism as shown in Fig. 3.19a, polycarbonate undergoes five stages of degradation. The first stage involves intramolecular exchange to produce cyclic oligomers

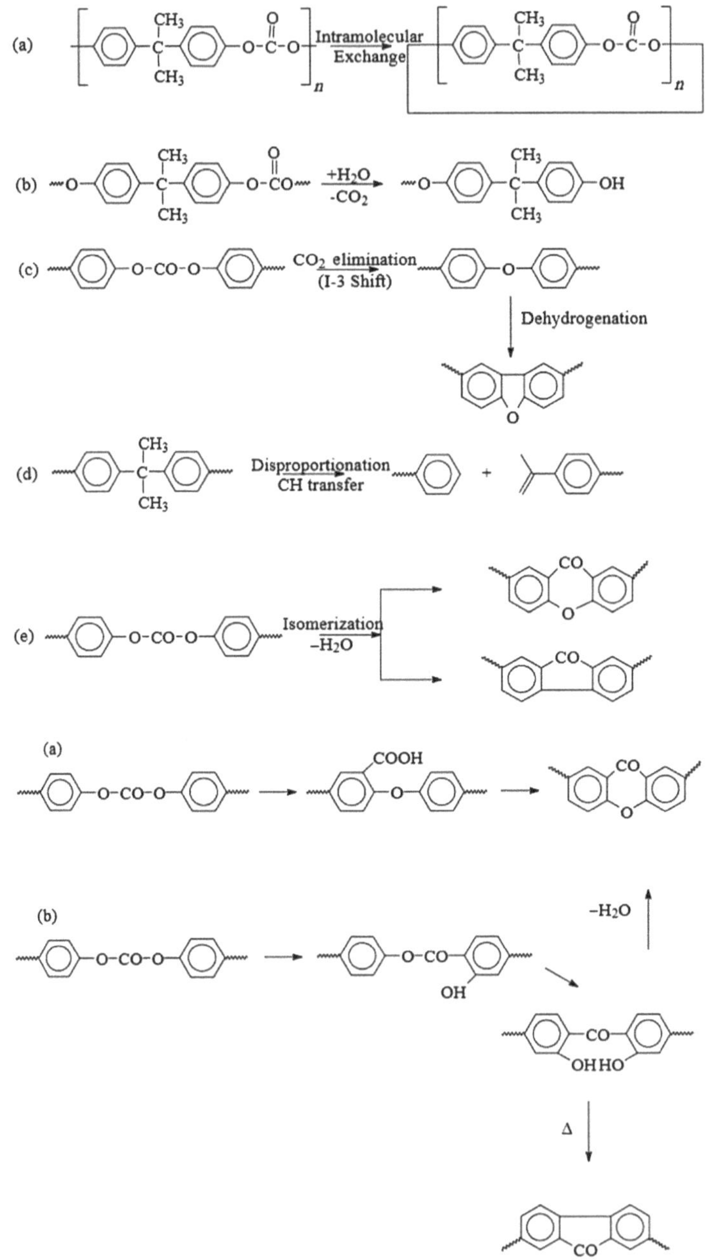

Figure 3.19 Montaudo and Puglisi's (a) polycarbonate degradation mechanism and (b) an alternative isomerization mechanism [44,46].

followed by hydrolytic cleavage of the carbonate group during the second stage. In the third stage, carbon dioxide is eliminated and then disproportioned in the fourth stage. The last stage involves isomerization of the polycarbonate group. Fig. 3.19b shows an alternative isomerization mechanism which yields xanthone units and fluorenone units [44,46].

3.5 Polyurethanes

Polyurethane (PU) can be defined as flexible to rigid depending on the chemical structures of the polyisocyanates and polyols. The chemical composition of PU precludes them from being classified as pure plastics and hence called as a mixed polymer.

The urethane group acts as a base unit of PU and forms the basis of this class of mixed polymer. It represents a small part of the macromolecule and some PU products do not contain a urethane group. Despite the lack of this base unit, all PUs are based on the composition of polyisocyanates. The polyisocyanates addition is distinct from polymerization and polycondensation for the production of synthetic polymers and this feature explains their versatility. Fig. 3.20 shows a generalized structure of PU [47,48].

In the figure, R represents a hydrocarbon containing the alcohol group, R2 is a hydrocarbon chain, and *n* is the number of repetitions. Diisocyanates are used in PU production reactions because they react with any compound that contains active hydrogen [47,48].

The degradation of PU involves two main mechanisms:

- Mineralization
- Oxidation

Mineralization or calcification is the mechanism in which crystalline inorganic substances are incorporated on the surface of the implanted material. This results in hardening, distortion, and mechanical degradation of the elastomer segment of PU. Oxidation is the mechanism that is

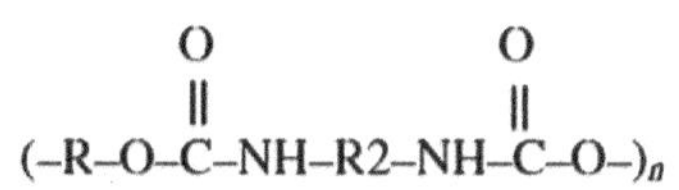

Figure 3.20 Generalized structure of PU [47,48].

induced by free-radicals breaking bonds and opening the elastomer up to attack by oxygen molecules. Oxidation is influenced by factors such as the flexibility, the type of soft segment, and the ability to interact with ions [49].

Fig. 3.21 shows a hydrolytic degradation mechanism of polyesters (a), PU (b) and polyuria (c). Because of ester groups in the soft segments in poly(ester urethanes), they are susceptible to hydrolytic degradation. However, poly(ether urethanes) are susceptible to oxidative degradation. It is important to note that ester linkages tend to hydrolyze significantly faster than the urethane linkages. Similarly, urea linkages hydrolyze much faster than urethane at slightly acidic conditions [50].

Fig. 3.22 shows mechanism for oxidative degradation by hydrogen peroxide in poly(ether urethanes), poly(carbonate urethanes), and aromatic PUs. Since many studies have shown that poly(ether urethanes) degrade by α-hydrogen abstraction adjacent to oxygen in polyethers, they are generally associated with poly(ether urethanes) [51,52].

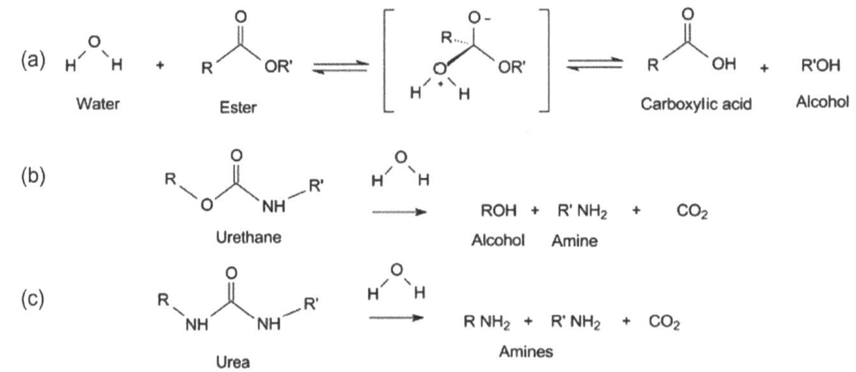

Figure 3.21 Hydrolytic degradation mechanism of polyesters (a), PU (b), and polyuria (c) [50].

(a) $\text{—CH}_2\text{-O-CH}_2\text{—} + {}^\bullet\text{OH} \longrightarrow \text{—CH}_2\text{-O-}\dot{\text{C}}\text{H—} + \text{H}_2\text{O}$

(b) $\text{—CH}_2\text{—OCOO-CH}_2\text{—} + {}^\bullet\text{OH} \longrightarrow \text{—CH}_2\text{—OCOO-}\dot{\text{C}}\text{H—} + \text{H}_2\text{O}$

(c) $\text{—Ph—NHCOO—CH}_2\text{—} + {}^\bullet\text{OH} \longrightarrow \text{—Ph—NHCOO—}\dot{\text{C}}\text{H—} + \text{H}_2\text{O}$

Figure 3.22 Mechanism for oxidative degradation by H_2O_2 in poly(ether urethanes) (a), poly(carbonate urethanes) (b), and aromatic PUs (c) [53–55].

In contrast, oxidative degradation mechanisms of polyester PUs have not been deeply researched and even much less for PU ureas. Recent studies have shown that PCL and polyester poly(urethane urea)s PCL have shown ester, urethane, and urea groups are susceptible to oxidative degradation [53–55].

3.6 Polyacetals

Polyoxymethylene (POM), also known as polyacetal, is a functional group or molecule containing the functional group in which carbon C is bonded with two –OR groups.

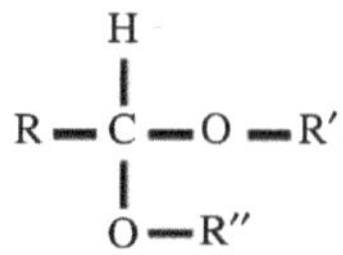

In principle, there are three main degradation mechanisms of POM:

1. Thermal oxidation
2. Photo-oxidation
3. Hydrolysis

3.6.1 Thermal Oxidation

Polyacetal has the tendency to rapidly depolymerize in the absence of oxygen while formaldehyde reacts spontaneously to form low molecular weight oligomers. The reaction starts at chain ends and depolymerization then proceeds sequentially along the backbone chain. Acetylation of the terminal hydroxyl group leads to end-capping of macromolecules.

Mechanism for thermal degradation of POM in the absence of oxygen is shown in Fig. 3.23. The decomposition of a stable fragment, C, with carbonyl groups at the chain end may occur during the second stage of the thermal degradation process [56].

In the presence of oxygen, the mechanism is shown in Fig. 3.24. This mechanism can readily account for the fact that the quantity of carbonyl groups formed in the decomposition in air is greater than that in nitrogen [57].

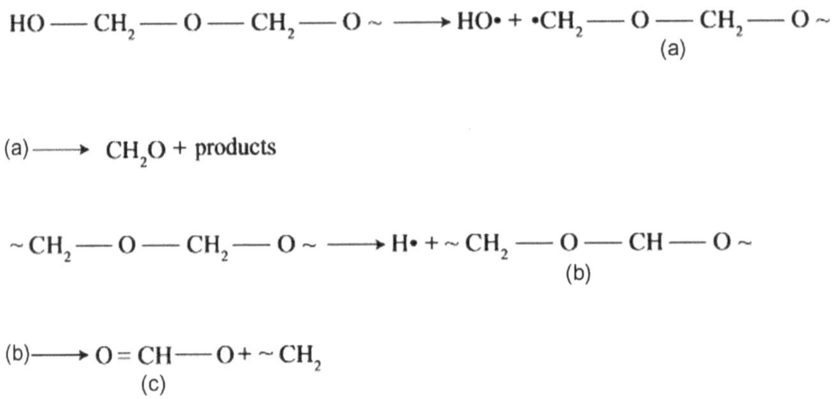

Figure 3.23 Mechanism for the thermal degradation of POM [56].

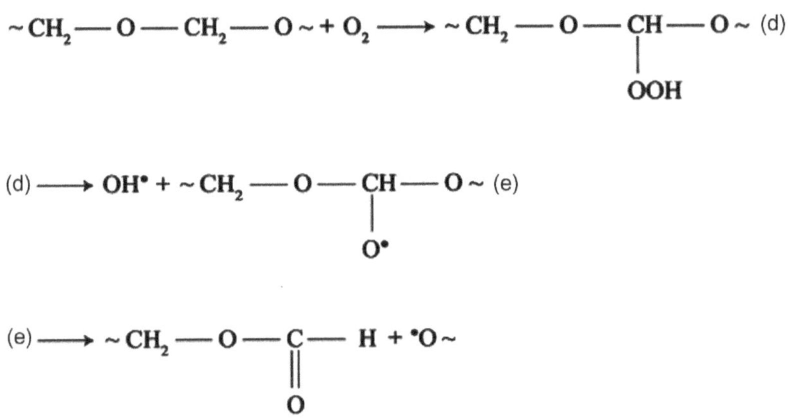

Figure 3.24 Depolymerization reaction of polyacetals [57].

3.6.2 Photo-Oxidation

Polyacetals are known to be oxidation-sensitive. The abstraction of hydrogen on the methylene group next to the oxygen atom yields a hydro peroxide group. Fig. 3.25 shows two different mechanisms of photo-oxidation of polyacetals [58]. Photolysis of the hydro peroxide group can happen in two ways: (a) cage reaction and (b) β-scission. Photolysis of the hydro peroxide in the cage leads to carbonate structures and water, and β-scission leads to aldehyde groups. Further oxidation results in the formation of acidic groups and, subsequently, to acid-catalyzed depolymerization of the polyacetal [58].

Polyacetal under the irradiation of UV light undergoes depolymerization as illustrated in Fig. 3.26 [58].

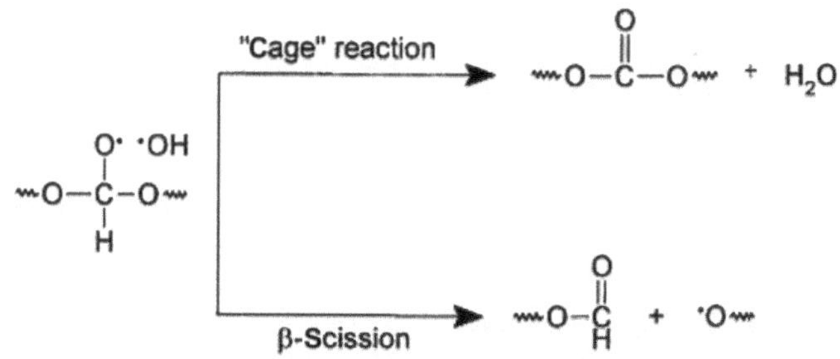

Figure 3.25 Different mechanisms of photo-oxidation of polyacetals [58].

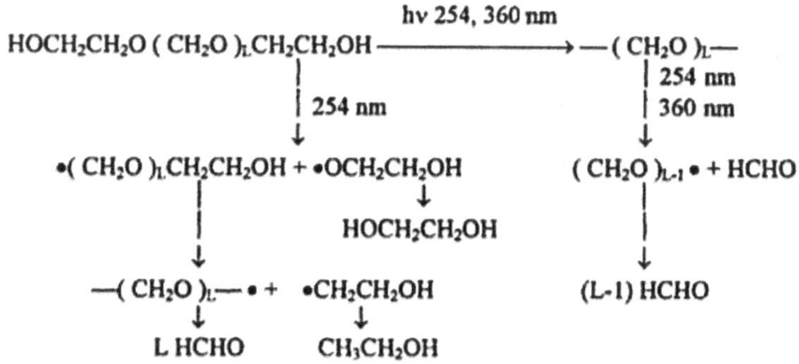

Figure 3.26 Polyacetal depolymerization mechanism under UV irradiation [58].

3.6.3 Hydrolysis

Acetals tend to acid hydrolysis due to the possible co-ordination of a proton with the unshared pair of electrons on the oxygen of the acetal group, resulting in the formation of the oxonium ion. The great instability of acetals in these conditions is due to the energy profit obtained by resonance with a carbonium ion form:

$$RO\overset{+}{C}H_2 \rightleftharpoons R\overset{+}{O}=CH_2$$

Coordination of hydroxyl with the acetal group is impossible, and this determines the high hydrolytic stability of acetals in alkaline media. The rate of hydrolysis of acetal is determined by the rate of carbonium ion formation, and the hydrolysis proceeds under specific acid catalysis as shown in Fig. 3.27 [58].

In addition, the limiting stage may appear to be either a monomolecular reaction corresponding to mechanism 1 or a bimolecular reaction which includes the water molecule in the intermediate complex corresponding to

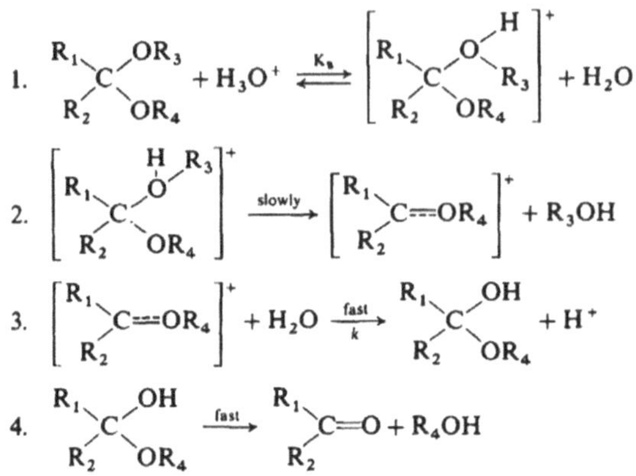

Figure 3.27 Hydrolysis of acetals via specific acid catalysis [58].

mechanism 2. The value of the rate constant of the limiting step of the hydrolysis process for each of these mechanisms can be calculated [58].

In the case of mechanism 1:

$$\log k = -H_0 + \log \frac{f_{BH^+} f_S}{f_B \times f_{\neq}} + \text{const}$$

In the case of mechanism 2:

$$\log k = \log C_{H^+} + \frac{f_S f_H a_{H_2O}}{f_{\neq}} + \text{const}$$

where $f_{\neq}, f_S, f_{H^+}, f_B, f_{BH^+}$ are activity coefficients of the transition complex ($\neq$), acetal (S), proton (H^+), indicator (B), and its conjugate acid (BH^+); a_{H_2O} is the activity of the water molecule, C_{H^+} is concentration of hydrated protons, and $-H_0$ is the Hammet acidity function.

References

[1] Degradation of polymers in nature, Dow Corning, Ref. no. 01-1112-01. <http://www.dowcorning.com/content/publishedlit/01-1112-01.pdf>.

[2] V. Batu, Applied Flow and Solute Transport Modeling in Aquifers: Fundamental Principles and Analytical and Numerical Method, CRC Press, Boca Raton, FL, 2006.

[3] P.N. Thanki, R.P. Singh, Photo-oxidative degradation of nylon 6,6 under accelerated weathering, Polymer 39 (25) (1998).

[4] C. Bastioli, Handbook of Biodegradable Polymers, iSmithers Rapra Publishing, Shawbury, Shrewsbury, Shropshire (UK), 2005, p. 5.

[5] R. Premraj, M. Doble, Biodegradation of polymers, Int. J. Biotechnol. 4 (2004) 186–193.
[6] M. Ramezani, D. Oliver, R. Kookana, G. Gill, C. Preston, Abiotic degradation (photodegradation and hydrolysis) of imidazolinone, J. Environ. Sci. Health B 43 (2) (2008) 105–112.
[7] B.Y. Wang, Environmental Biodegradation Research Focus, Nova Publishers, New York, 2007.
[8] S. Kaur, Mechanism of aerobic and anaerobic biodegradation, M.Sc. Thesis, 2014.
[9] C.H. Banford, C.F.H. Tipper, Comprehensive Chemical Kinetics, vol. 14, Degradation of Polymers, Elsevier, Amsterdam, 1972.
[10] M.M. Reddy, R.K. Gupta, R.K. Gupta, Abiotic oxidation studies of oxo-biodegradable polyethylene, J. Polym. Environ. 16 (1) (2008) 27–34.
[11] S.A. Ashter, Thermoforming of Single and Multilayer Laminates: Plastics Films Technologies, Testing, and Applications, Elsevier, Waltham, MA, USA, 2013 (Chapter 3).
[12] M. Eastman, Materials world modules, The University of Texas at El Paso. <http://materialsworld.utep.edu/Modules/FoodPackaging/Food%20Packaging%20intro%20cont/Biodegradable%20Polymers/biodegradable%20polymers.htm>.
[13] T.C. Montie, Pseudomonas, Biotechnology Handbooks, vol. 10, Plenum Press, New York, 1998.
[14] A.-C. Albertsson, Degradable Aliphatic Polyesters, Springer, New York, 2003.
[15] F. Von Burkersroda, L. Schedl, A. Göpferich, Why degradable polymers undergo surface erosion or bulk erosion, Biomaterials 23 (2002) 4221–4231.
[16] X.S. Wu, in: D.L. Wise, D.J. Trantolo, D.E. Altobelli, M.J. Yaszemski, J.D. Gresser, E.R. Schwarz (Eds.), Encyclopedic Handbook of Biomaterials and Bioengineering, Marcel Dekker, New York, 1995, pp. 1015–1054.
[17] A. Göpferich, J. Tessmar, Polyanhydride degradation and erosion, Adv. Drug Deliv. Rev. 54 (2002) 911–931.
[18] Y. Sugimura, S. Tsuge, Studies on thermal degradation of aromatic polyesters by pyrolysis-gas chromatography, J. Chromatogr. Sci. 17 (1979) 269–272.
[19] I. Luderwald, H. Urrutia, Strukturuntersuchung von polyester durch direkter Abbau im massenspectrometer, vol. 1, Macromol. Chem. 177 (1976) 2079.
[20] I. Luderwald, H. Urrutia, Direct pyrolysis of aromatic and aliphatic polyesters in the mass spectrometer, in: C.E.R. Jones, C.A. Cramers (Eds.), Analytical Pyrolysis, Elsevier, Amsterdam, 1977, p. 139.
[21] V. Passalacqua, F. Pitati, V. Zamboni, B. Fortunato, P. Manaresi, Thermal degradation of poly(butylene terephthalate), Polymer 17 (1976) 1044.
[22] B.C. Ho, Y.D. Lee, W.K. Chin, Thermal degradation of polymethacrylic acid, J. Polym. Sci. Part A Polym. Chem. 30 (11) (1992) 2389–2397.

[23] H.G. Schild, Thermal degradation of poly(methacrylic acid): further studies applying TGA/FTIR, J. Polym. Sci. Part A Polym. Chem. 31 (9) (1993) 2403–2405.
[24] M. Ferriol, A. Gentilhomme, M. Cochez, N. Oget, J.L. Mieloszynski, Thermal degradation of poly(methyl methacrylate) (PMMA): modelling of DTG and TG curves, Polym. Degrad. Stab. 79 (2) (2003) 271–281.
[25] Z. Ping, W. Linbo, L. Bo- Geng, Thermal stability of aromatic polyesters prepared from diphenolic acid and its esters, Polym. Degrad. Stab. 94 (2009) 1261–1266.
[26] D.N. Bikiaris, K. Chrissafis, K.M. Paraskevopoulos, K.S. Triantafyllidis, E.V. Antonakou, Investigation of thermal degradation mechanism of an aliphatic polyester using pyrolysis-gas chromatography-mass spectrometry and a kinetic study of the effect of the amount of polymerization catalyst, Polym. Degrad. Stab. 92 (2007) 525–536.
[27] K. Pielichowski, J. Njuguna, Thermal Degradation of Polymeric Materials, iSmithers Rapra Publishing, Shawbury, Shrewsbury, Shropshire (UK), 2005, pp. 73–75.
[28] J.F. Rabek, Polymer Photodegradation: Mechanisms and Experimental Methods, Springer, New York, 1995.
[29] G.S. Egerton, J. Text. Inst. 39 (1948) T293.
[30] G.S. Egerton, J. Soc. Dy. Col. 65 (1949) 764.
[31] R.F. Moore, Photochemical degradation of polyamides and related model *n*-alkylamides, Polym. J. 4 (1963) 493.
[32] P. Jan, P. Klemchuk, Oxidation Inhibition in Organic Materials, vol. 1, CRC Press, Boca Raton, FL, 1989.
[33] M.N. Grigg, Thermo-oxidative Degradation of Polyamide 6, Doctoral Thesis, School of Physical and Chemical Sciences, Queensland University of Technology, 2006.
[34] T. Karstens, V. Rossbach, Macromol. Chem. 190 (1989) 3033.
[35] V. Rossbach, T. Karstens, Chemiefasern/Textilind. 40 (1990) E44.
[36] T. Yoshioka, G. Grause, Hydrolysis of polyesters and polycarbonates, Progress in Polymer Degradation and Stability Research, Nova Science Publishers, New York, 2008 (Chapter 1).
[37] A. Davis, J.H. Golden, J. Gas Chrom. 5 (1967) 81.
[38] A. Davis, J.H. Golden, J. Chem. Soc. Part B. Phys. Org. 1-7 (1968) 47.
[39] A. Davis, J.H. Golden, J. Macromol. Sci.-Revs. Macromol. Chem. C3 (1969) 49.
[40] L.H. Lee, J. Polym. Sci. Part A 2 (1964) 2859.
[41] S. Kuroda, K. Terauchi, K. Nogami, I. Mita, Eur. Polym. J. 25 (1989) 1.
[42] I.C. McNeill, A. Rincon, Polym. Degrad. Stab. 31 (1991) 163.
[43] I.C. McNeill, A. Rincon, Polym. Degrad. Stab. 39 (1993) 13.
[44] G. Montaudo, C. Puglisi, in: N. Grassie (Ed.), Developments in Polymer Degradation, Elsevier, New York, 1987, p. 35.
[45] G. Montaudo, C. Puglisi, Polym. Degrad. Stab. 37 (1992) 91.

[46] C. Puglisi, L. Sturiale, G. Montaudo, Macromolecules 32 (1999) 2194.
[47] B.A. Dombrow, Polyurethanes, Reinhold Publishing Corporation, New York, 1957.
[48] S. Dumitriu, Polymeric Biomaterials, Revised and Expanded, CRC Press, Boca Raton, FL, 2002.
[49] J.V. Cauch-Rodriguez, L.H. Chan-Chan, F. Hernandez-Sanchez, J.M. Cervantes-Uc, Degradation of polyurethanes for cardiovascular applications, Advances in Biomaterials Science and Biomedical Applications, InTech Publishers, 2013 (Chapter 3).
[50] E.M. Christenson, M. Dadsetan, M. Wiggins, J.M. Anderson, A. Hiltner, J. Biomed. Mater. Res. Part A 69 (3) (2004) 407–416.
[51] X. Xie, R. Wang, J. Li, L. Luo, D. Wen, Y. Zhong, et al., J. Biomed. Mater. Res. Part B 89B (1) (2009) 223–241.
[52] M.A. Sabino, Polym. Degrad. Stab. 92 (6) (2007) 986–996.
[53] D. Sarkar, S.T. Lopina, Polym. Degrad. Stab. 92 (11) (2007) 1994–2004.
[54] A.E. Hafeman, K.J. Zienkiewicz, A.L. Zachman, H.J. Sung, L.B. Nanney, J.M. Davidson, et al., Biomaterials 32 (2) (2011) 419–429.
[55] W.L. Hawkins, Polymer Degradation Stabilization, Springer-Verlag, Berlin, Heidelberg, 1984.
[56] T.R. Crompton, Thermo-Oxidative Degradation of Polymers, Smithers Rapra Publishing, 2010.
[57] G.O. Shonaike, G.P. Simon, Polymer Blends and Alloys, Marcel Dekker, New York, 1999.
[58] V.V. Pchelintsev, A.Y. Sokolov, G.E. Zaikov, Kinetic principles and mechanisms of hydrolytic degradation of mono- and polyacetals—a review, Polym. Degrad. Stab. 21 (1988) 286.

4 Fundamentals on Biodegradability

4.1 Introduction

4.1.1 What Is Biodegradability?

There has been a significant increase in the use of biomaterials within the last decade. However, clarity in defining different terminologies such as biodegradation and composting has been a challenge [1]. Both of these are natural processes in which organic material is converted into environmentally accepted molecules such as water, carbon dioxide, minerals, and biomass should be clearly defined. The biodegradability of a polymer provides material with additional properties that make it useful in a variety of applications. One such example is use of agricultural biodegradable mulch films. Their biodegradable property makes them useful because they are able to decompose and mix with soil in time. Hence, it is not surprising that this concept of using biodegradable polymers has become of major interest during recent years [2].

Fig. 4.1 shows biodegradable plastic forks, knives, and spoons disintegrate into the soil. Their ability to be broken down into simpler substances by bacteria is termed as biodegradability.

Substances that are capable of biodegrading are referred to as bioactive. Real-life examples of such bioactive substances are bread, vegetables, and fruits which break down to become simple sugars. Some other materials, like leftover hot dogs, break down into carbon, oxygen, and water. However, it is important to note that not all substances are bioactive. Most plastics, for example, are not bioactive. Biodegradation occurs aerobically or anaerobically. Under aerobic conditions, oxygen helps to break molecules apart, and substances break down much faster [3].

Some materials are broken down into organic matter much faster than others. They are often referred to as biodegradable. How long a material will stay in its form is determined by measuring material's biodegradability [4]. The biodegradability is generally initiated by the decomposition of any object buried in the ground. This object is slowly broken down into smaller elements when it gets exposed to microorganisms and bacteria. For some objects, the breakdown starts to become visible within hours and in some cases it may take years to show up.

Introduction to Bioplastics Engineering.
DOI: http://dx.doi.org/10.1016/B978-0-323-39396-6.00004-X

Figure 4.1 An example of biodegradability [3].

This has brought up a very important question about how much non-biodegradable material ends up in trash. So, there has been a big push for more biodegradable products in order to answer this question.

Biodegradability also plays an important role in stability of the environment. Although some of the biodegradable materials break down quickly, some of the materials such as plastics can last for decades. These nondegradable materials end up in landfills pushing them over their capacity [4].

Materials that are generated either from plant or animal material have a high rate of biodegradability. This is because earth's microorganisms are capable of breaking down such material in a natural environment. However, materials that are made by humans not necessarily contains naturally occurring microorganism to break the material down [4]. This has led to the use of two important terminologies—biodegradation and composting. These two terms are used interchangeably; however, meaning of these terms is not clearly defined. This has created obscurities in their meanings and should be clearly defined. [3].

In contrast to biodegradability, compostability is the biodegradability of a material that uses compost medium. This biodegradation happens as a result of biotic degradation primarily, an enzymatic action of microorganism. This results in CO_2, new biomass, and water when the degradation happens under aerobic conditions or methane under anaerobic conditions [5]. The end result of this compostable biodegradation is that the object decomposes and thus blends in with the earth soil similar to what is shown in Fig. 4.2 [4,6].

Figure 4.2 An example of compostability [1,4,6].

During the new product development cycle, biodegradability is given primary consideration. More and more manufacturers are moving toward the use of biodegradable products. Biodegradable polymers have found their use in medical device industry. Synthesis of new biodegradable plastic has been created to allow faster biodegradation by lowering the molecular weight of the components [4,7]. Biodegradable medical implants are one of the fastest growing areas in the orthopedics industry and is expected to grow biodegradable polymers that are used in a range of surgical applications, including joint replacements, fracture fixation plates, bone defect fillers, artificial tendons and ligaments, and bone cements. Other medical applications include cardiovascular, ophthalmic, sutures, burn and wound dressings, drug delivery systems, cochlear implants, and dental implants [8,9].

Packaging is another area where the use of biodegradable products is on the rise. Manufacturers are moving toward packaging that is made of easily recyclable and quickly degrading paper, as opposed to the previously popular and slow decaying Styrofoam or plastic packing material.

4.1.2 Biodegradability Standards

There are existing standard tests that can measure the mechanism of biodegradation process via the enzymatic action of microorganisms over time. By measuring the biodegradation process, these standards can reflect available disposal conditions. According to the European Standard EN 13432-2000, there are different media to analyze

Table 4.1 Standardization Nomenclature [5]

Standard	Description
ISO	International Organization for Standardization
CEN	European Committee for Standardization
AFNOR	Organization of the French Standardization System
JIS	Japanese Industrial Standards
DIN	Deutsches Institut für Normung e. V. (German Institute for Standardization)
BS	British Standards
ASTM	American Society for Testing and Materials

biodegradability. These media could be liquid, inert, or compostable. Different standards, as shown in Table 4.1, utilize different composting conditions to determine the compostability level [5].

Table 4.2 shows biodegradation terminology that is published by national and international standards and organizations [2,10]. Each standard lists the meaning of biodegradation in a standard which can differ from one standard to another. For example, ISO defines biodegradable plastics as a chemical change of the material by microorganisms while CEN and Deutsches Institut für Normung e. V. (DIN), in contrast, demand in their definitions the conversion of plastics into microbial metabolic products [2]. Other definitions such as inherent biodegradability or ultimate biodegradability are adapted from the area of degradation of low-molecular-weight chemicals, but these can also be applied to polymers. Although, the definitions shown in Table 4.2 do not specifically apply to any environment or time frame, they must be performed in accordance to the corresponding standards [2].

Fig. 4.3 shows schematic view of different standards [11]. These standards are grouped into three levels:

1. Global — eg, ISO
2. Regional — eg, CEN
3. National — eg, AFNOR, JIS, DIN, BS, ASTM

The global standard such as ISO is accepted worldwide. Regionally-accepted standard such as CEN is more commonly used in Europe. National standards are the ones which apply to local and national locations. For example, AFNOR is primarily used in France, JIS in Japan, DIN in Germany, BS in United Kingdom, and ASTM in the United States.

Table 4.2 Biodegradation Terminology Used in Standards [2,10]

Standard	Definition
DIN	*Biodegradable plastics* A plastic material is called biodegradable if all its organic compounds undergo a complete biodegradation process. Environmental conditions and rates of biodegradation are to be determined by standardized test methods *Biodegradation* Biodegradation is a process, caused by biological activity, which leads under change of the chemical structure to naturally occurring metabolic products
ASTM sub-committee D20-96	*Biodegradable plastics* A degradable plastic in which the degradation results from the action of naturally occurring microorganisms such as bacteria, fungi, and algae
Japanese Biodegradable Plastics Society	*Biodegradable plastics* Polymeric materials which are changed into lower molecular weight compounds where at least one step in the degradation process is through metabolism in the presence of naturally occurring organisms.
ISO 472	*Biodegradable plastics* A plastic designed to undergo a significant change in its chemical structure under specific environmental conditions resulting in a loss of some properties that may vary as measured by standard test methods appropriate to the plastic and the application in a period of time that determines its classification. The change in the chemical structure results from the action of naturally occurring microorganisms

Table 4.2 Biodegradation Terminology Used in Standards [2,10]—cont'd

Standard	Definition
CEN	*Biodegradable plastics* A degradable material in which the degradation results from the action of microorganisms and ultimately the material is converted to water, carbon dioxide and/or methane, and a new cell biomass *Biodegradation* Biodegradation is a degradation caused by biological activity, especially by enzymatic action, leading to a significant change in the chemical structure of a material *Inherent biodegradability* The potential of a material to be biodegraded, established under laboratory conditions *Ultimate biodegradability* The breakdown of an organic chemical compound by microorganisms in the presence of oxygen to biodegradability carbon dioxide, water and mineral salts of any other elements present (mineralization), and new biomass or in the absence of oxygen to carbon dioxide, methane, mineral salts, and new biomass *Compostability* Compostability is a property of a packaging to be biodegraded in a composting process. To claim compostability, it must have been demonstrated that a packaging can be biodegraded in a composting system as can be shown by standard methods. The end product must meet the relevant compost quality criteria

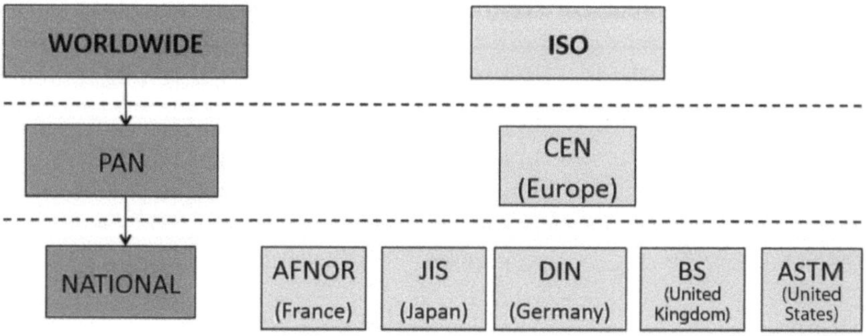

Figure 4.3 Levels of standardization [11].

	WORLDWIDE	EUROPE	US	AUSTRALIA
PLASTICS	ISO 17088	EN 14995	ASTM D6400	AS 4736
PACKAGING	ISO 18606	EN 13432		
PAPER COATING			ASTM D6868	

Figure 4.4 Standard specifications: tests and criteria [11].

All of these standards differ in the way the standard specifications are set which includes testing and criteria. Fig. 4.4 shows an example of such difference based on the geographical location for a given application [11].

4.1.3 Biodegradability Testing

When evaluating biodegradability of plastics, there has always been a concern regarding the type of tests needed to be performed and what kind of conclusions can be drawn from these tests. In essence, these tests are used to predict biodegradability of a test material in its natural environment. That is why these tests should be able to simulate the environments to a level of accuracy. The better the level of simulation, accurate the prediction is and subsequently, select the kind of testing to be performed [12]. Table 4.3 shows some of tests performed by different standard organization [13].

Table 4.3 Tests Under Different Standardization Procedure [13]

ASTM testing methods	ISO testing methods	OECD testing methods
E1720	ISO 9439	301A
D5210	ISO 14593	301B
D5338	ISO 14852	301C
D5511	ISO 14855	301D
D5864	ISO 15985	301E
D5988		301F
D6400		310
D6954		311

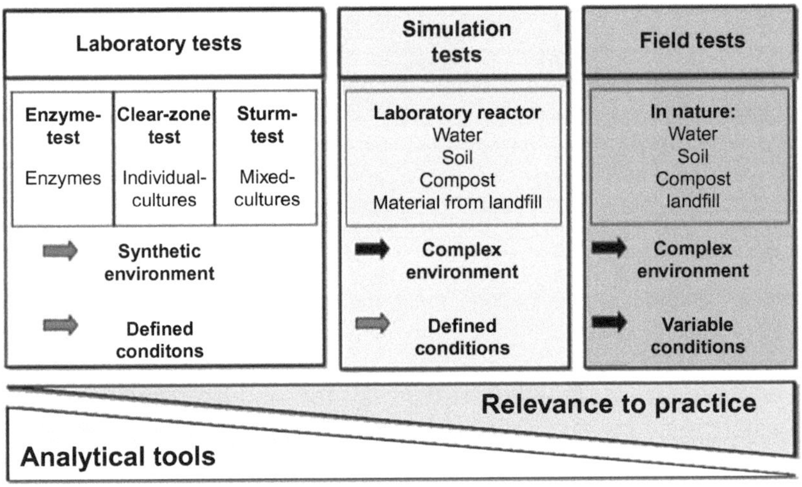

Figure 4.5 Testing overview for biodegradable plastics [2].

Fig. 4.5 shows overview of different testing performed on biodegradable plastics. In principle, tests can be subdivided into three categories [2]:

1. Laboratory tests
2. Simulation tests
3. Field tests

A laboratory test is an analytical tool to evaluate biodegradability of a material in lab setting. They provide synthetic environment and tests are performed under defined conditions. Fig. 4.6 shows an example of the laboratory composting system and vessel [14].

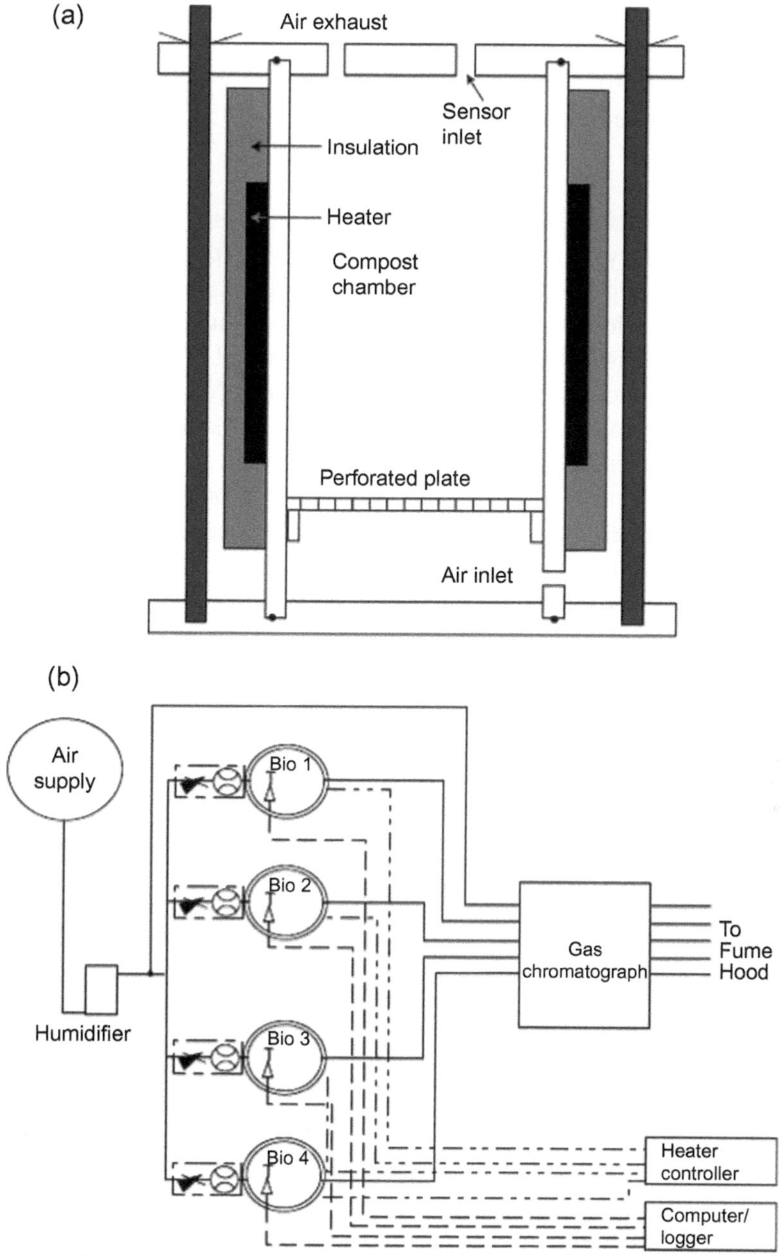

Figure 4.6 Example of (a) the laboratory composting vessel and (b) the laboratory composting system [14].

There are three different types of tests that come under this category [2]:

- Enzyme test
- Clear-zone test
- Sturm test

Simulation tests are performed by creating an artificial environment such as laboratory reactor to mimic a natural biodegradable environment. This testing approach is not just mere analytical tool but has more practical relevance. In this test, water, soil, and compost material from the landfill is loaded in the reactor and mixed. The mix is let to sit in the reactor depending on the compost material for it to decompose if it was in its natural environment. This process works in a complex environment with defined conditions.

The last test is the one which is most relevant to practice. This testing is performed at the test site—in nature and hence it is referred to as field test. This test works in a complex environment and conditions can be varied.

4.2 Measurement of Biodegradability of Polymers

Biodegradability of a polymer depends on the rate at which degrades. For example, an apple core will degrade within few weeks; however, the plastic toy can take decades to completely disappear. The three key factors that help microorganisms to degrade a material are light, water, and oxygen. In order to work more effectively, these three factors needs to be careful controlled. In addition to these factors, temperature is also an important factor in determining the rate of biodegradation. This is because microorganisms tend to reproduce faster in warmer conditions [15].

Biodegradability is the key to determine measurement guideline. There are certain tests to measure the biodegradability of a material. In one such test procedure, it requires introducing a sample of the organic material into a healthy bacterial suspension under controlled conditions and measuring the time or rate at which a suitable end point is achieved. Microorganisms use the carbon substrates to extract chemical energy that drives their life processes by aerobic oxidation of glucose and other readily utilizable C-substrates:

$$C\text{-substrate} + 6O_2 \rightarrow 6CO_2 + 6H_2O, \ \Delta G^0 = -686\ \text{kcal/mol}$$
$$(CH_2O)_x; \quad x = 6$$

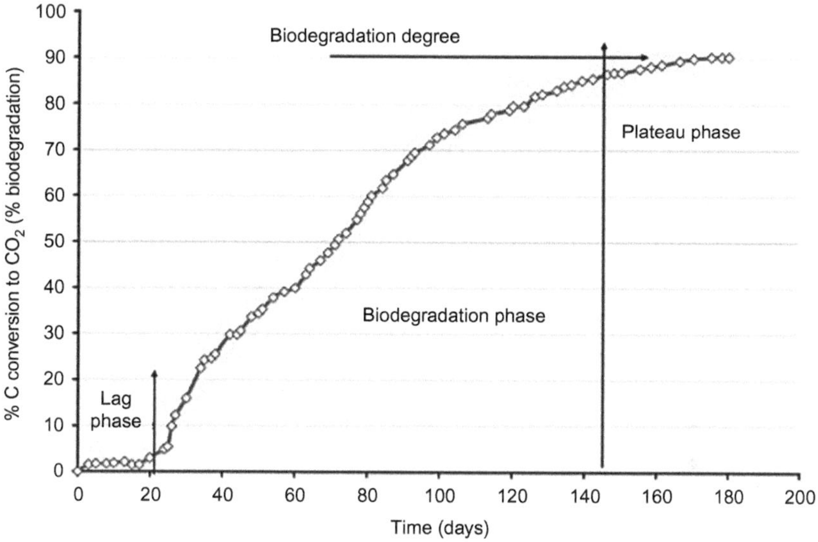

Figure 4.7 Biodegradation curve [11,15].

By measuring the rate and amount of carbon dioxide evolved in the process, it directly relates to the amount and rate of biodegradation of the C-polymer. This creates a foundation for different international standards for measuring biodegradability of plastics. With this procedure, one can measure the rate and extent of biodegradation of plastics by using it as the sole added carbon source in a test system containing a microbial-rich, matrix-like compost in the presence of air and under optimal temperature conditions, preferably at 580°C [15].

Percent biodegradation with respect to time can be plotted to draw out a curve as shown in Fig. 4.7 [11,15]. It is also known as biodegradation curve. As seen from Fig. 4.7, the percent biodegradation increases with time. The lag phase at the start sees the microbial population adapts to the available test carbon substrate. Then, the biodegradation phase during which the microbes begin to utilize carbon substrate for its cellular life processes. Finally, the area where percent biodegradation reaches a plateau which means utilization of the substrate is largely complete [16].

Requirements of the international standards for complete biodegradation under composting conditions are as follows:

1. Conversion to carbon dioxide, water, and biomass via microbial assimilation of the test polymer material in powder, film, or granule form.
2. 90% conversion of the carbon in the test polymer to carbon dioxide.

3. Same rate of biodegradation as natural materials—leaves, paper, grass, and food scraps
4. Time — 180 days or less. This can be extended to 365 if radiolabeled polymer is used and radiolabeled evolved carbon dioxide is measured.

There are two more requirements which are of importance:

1. Disintegration — Less than 10% of test material on a 2-mm sieve using the test polymer material in the shape and thickness identical to the product's final intended use according to ISO standards, ISO 16929 and ISO 20200.
2. Safety — The resultant compost should have no impacts on plants, using OECD Guide.

According to ASTM, the term biodegradable plastic is defined as a degradable plastic in which degradation results from the action of naturally occurring microorganisms such as bacteria, fungi, and algae [15]. These microorganisms are responsible for plastic biodegradation by using it as a food source. During aerobic biodegradation, carbon in polymer molecules is converted by microorganisms into water, biomass, carbon residues, and carbon dioxide. Therefore, the degree of aerobic biodegradation can be determined by measuring the total amount of carbon dioxide gas evolved from the polymer. Another term, percent mineralization has also been used to refer degradation. It is the proportion of cumulative carbon dioxide generated by the sample tested to the theoretical carbon dioxide content of the material and is calculated from chemical composition according to ASTM D5338 and ISO 14855 [17,18].

There are two main factors affecting the rate of polymer biodegradation. These factors are as follows[5]:

1. Exposure conditions
2. Polymer

The main exposure conditions are water or moisture, pH, temperature, and aerobic or anaerobic conditions. The polymer factors include its chemical structure, morphology, and molecular weight. It is critical to be able to assess all these factors in order to obtain reliable and reproducible data.

Table 4.4 shows a breakdown chart which shows how long some common items take to break down if left in the environment [13].

Table 4.4 Breakdown Chart [13]

Vegetables	**5 days–1 month**
Paper	2–5 months
Cotton T-shirt	6 months
Orange peels	6 months
Tree leaves	1 year
Wool socks	1–5 years
Plastic-coated paper milk cartons	5 years
Leather shoes	25–40 years
Nylon fabric	30–40 years
Tin cans	50–100 years
Aluminum cans	80–100 years
Glass bottles	1 million years
Styrofoam cup	500 years to forever
Plastic bags	500 years to forever

4.2.1 Biodegradability Measuring Devices

Biodegradability measurement devices will depend on the kind of test is utilized. Some of the major tests are as follows:

1. Plate test system
2. Respirometric system
3. Gas evolution test system

4.2.1.1 Plate Test System

In a typical plate test system, the plates are composed of two solid medium layers. The lower layer contains mineral salts for bacteria and fungi, respectively. The upper layer contains the polymer suspension. Typically, inoculation techniques are used to test the bacteria. One of the inoculation approaches is by using a dilution technique. Each milliliter of inoculum, containing one to ten bacteria according to the most probable number calculation, is added, and mixed with the liquid polymer suspension agar at a temperature between 45°C and 50°C, just before pouring it onto the mineral agar layer. The second approach is a plate-puncture technique, which is not only easier to handle but also allows higher reproducibility [19].

4.2.1.2 Respirometric System

A respirometric system or respirometer is a device that monitors respiration activity of living organisms either by focusing on oxygen consumption (O_2), carbon dioxide (CO_2) evolution or both. Respirometer consists of three major components:

1. An air supply
2. An air-tight closed vessel called the bioreactor which contains the living test organisms
3. A measuring device to quantify the amount of oxygen uptake or carbon dioxide release

4.2.1.2.1 MSU Respirometric Systems

Fig. 4.8 shows schematics drawing of cumulative measurement respirometric system [20]. Based on ASTM D 5338 [17], respirometric systems can be categorized by measuring technique. They are generally divided into two types:

1. Cumulative measurement respirometric
2. Direct measurement Respirometric

Acid–base titration is normally used in the cumulative measurement technique. The evolved carbon dioxide gas is trapped in a sodium hydroxide or barium hydroxide solution. Then, the trap solution is titrated using a known concentration of hydrochloric acid [20].

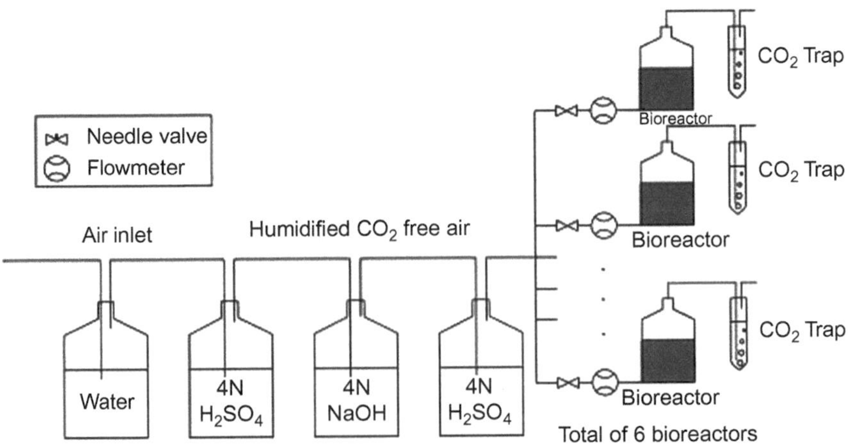

Figure 4.8 Schematic drawing of cumulative measurement respirometric system [20].

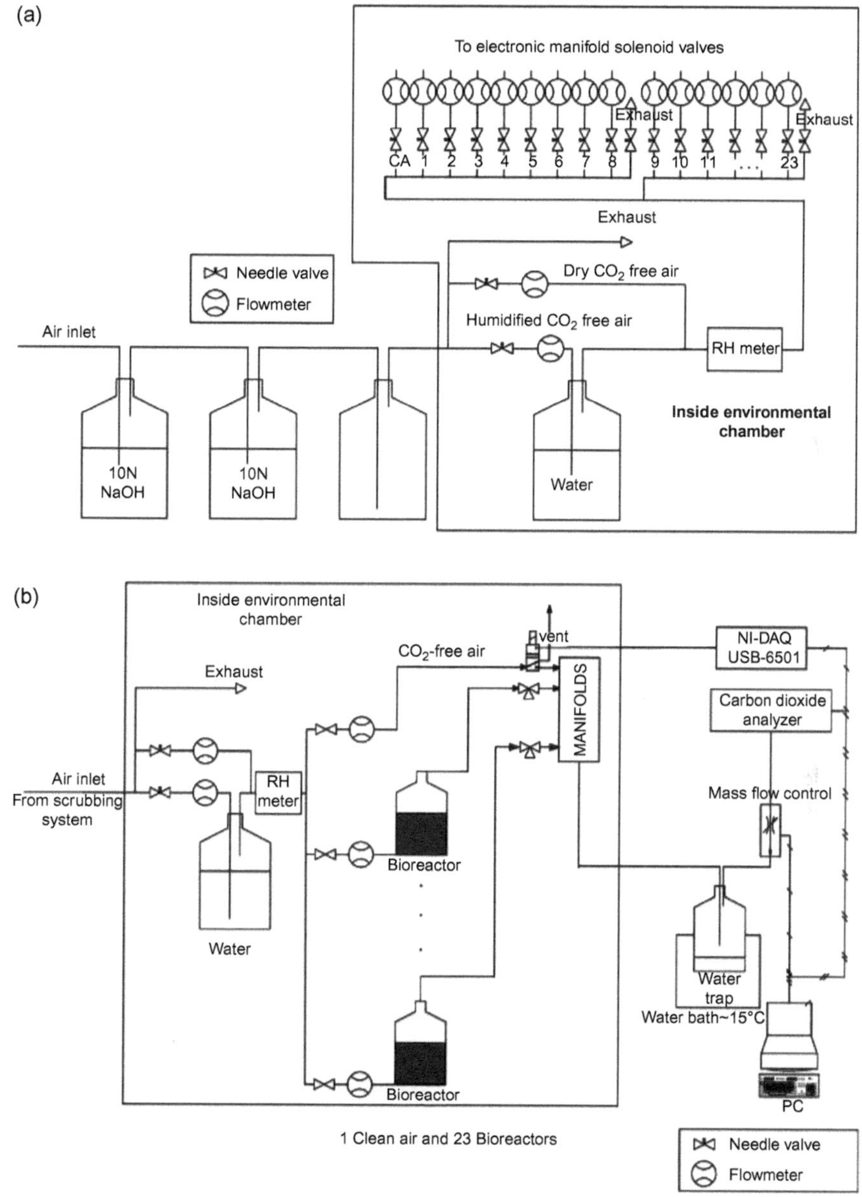

Figure 4.9 Schematic drawing of direct measurement respirometric system, (a) flowmeters and humidity control; (b) respirometric system components [20,21].

A direct measurement respirometer was built at the Michigan State University School of Packaging as shown in Fig. 4.9a and b [20,21]. For the direct measurement technique, the concentration of carbon dioxide is measured directly from the outlet of the bioreactor using gas chromatography, or by an inline infrared gas analyzer. For this method, the concentration of the exhaust gas of each bioreactor has to be measured at least once every 6 h [3].

This instrument has the capacity to simulate various testing conditions, from 0oC to an elevated phase temperature of 70°C, and can be supplied with either dry or humidified air. The system was designed to operate with up to 23 bioreactors of varying volumes (ie, 1, 2, 3, 5 L) at the same time, with the exhaust air cycled for measurement. It is possible to measure carbon dioxide evolved by setting the software in an automatic mode where each bioreactor will be tested at predetermined intervals throughout the duration of the experiment.

4.2.1.2.2 UMIC Lab I

The UMIC Lab I is an automatic measurement system that monitors factors such as carbon dioxide, temperature, relative humidity, and air flow. The device can hold up to 18 vessels of 3 L total volume and all the connections are air tight. The use of a closed bed reduces energy requirement to maintain a stable temperature. Each vessel is made independent from the other vessels by its own pressure regulator and, in the case of ISO 14852, with its own stirring plate. Gas switching system, sensor array, and pressure and flow regulation systems are installed inside the incubator with the vessels avoiding by this way any condensation problems. The conditions used in this equipment are designed to determine the potential biodegradability of polymers or give an indication of their biodegradability in natural environments [22].

4.2.1.3 Gas-Evolution Test System

Gas-evolution test system measures the gas evolution rate and volumes from exothermic decomposition of reaction masses at atmospheric pressure. This test system performs screening for unstable reactor contents and provides an input on whether the decomposing mass would be able to vent. This system consists of a glass U tube, which is partly filled with oil and equipped with a gas venting port, gas inlet port, and pressure port. A bottom runoff valve is also included for oil removal [23].

The volumetric gas evolved is measured by the electronic signal which is proportional to the gas evolution rate and total volume measured. When gas enters the U tube via the inlet port, gas pressure displaces the oil medium in one leg of the tube. This causes a change in head pressure as read by the pressure transducer. Pressure changes are monitored by the signal conditioning electronics and when a preset pressure is reached, as set by the user, the solenoid valve opens up causing accumulated gas to vent [23].

The number of gas vents to atmosphere is counted and the stored count value is displayed on the front panel of the instrument. Therefore, the number of gas vents multiplied by one collected volume equates to the total volume of gas evolved. A chart recorder or computer data logging system may be employed to record the gas evolved for each cycle of operation.

4.3 Factors Affecting Biodegradability

There are various methods that can be used to estimate the biodegradability of plastics [24–27]. It is always desirable to estimate biodegradability of plastics in its exposed natural environment such as in soil and sea water. However, by using this approach, it is not expected to have reproducible results. Similarly, a method which uses the specific enzyme [27–29] or bacteria [27,30] degrades the plastics and is able to measure the biodegradability quickly with good reproducibility. However, it is difficult to correlate the result with the biodegradability in natural environment. Quick and reproducible estimation of the biodegradability correlating to the biodegradability in natural environment is required to laboratory biodegradability test method.

The biodegradability of a plastic is often associated with its physical properties. The mechanism of biodegradation gets directly influenced by chemical and physical properties of plastic. The other factors that directly contribute toward biodegradability are as follows:

1. Surface condition of the plastic
2. First order structure such as chemical structure, molecular weight, and molecular weight distribution
3. High-order structures such as glass transition temperature, melting temperature, crystallinity, and crystal structure

The molecular weight is also important for the biodegradability because it determines many physical properties of the polymer. Increasing the molecular weight of the polymer decreased its degradability. Moreover, the morphology of polymers greatly affects their rates of biodegradation. The degree of crystallinity is also a crucial

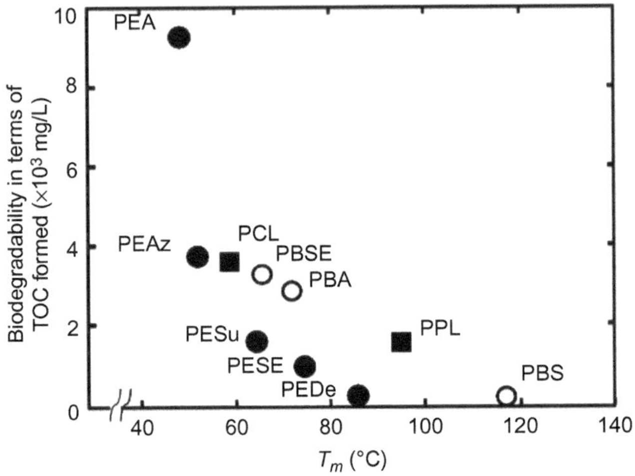

Figure 4.10 Relationship between T_m and biodegradability of polyesters [33].

factor affecting biodegradability, since enzymes mainly attack the amorphous domains of a polymer. The molecules in the amorphous region are loosely packed, and thus make it more susceptible to degradation. Fig. 4.10 shows a relationship between the melting temperature (T_{m}) and the biodegradability of polyesters. It can be inferred from Fig. 4.10 that polyesters have a strong inverse effect on the enzymatic degradation of polymers, the higher the melting temperature the lower the biodegradation of the polymer [31,32].

Melting temperature, T_{m} is represented by the following formula:

$$T_{\mathrm{m}} = \frac{\Delta H}{\Delta S}$$

where ΔH was the change of enthalpy in melting and ΔS is the change of entropy in melting. It is well-known that the interactions among polymer chains mainly affect the ΔH value and that the internal rotation energies corresponding to the rigidity (the flexibility) of the polymer molecule remarkably affect the ΔS value.

References

[1] Biodegradability and Compostability, <http://www.bio4expo.com/eng/insights/regulations/biodegradability-and-compostability>.

[2] R.-J. Muller, Biodegradability of Polymers: Regulations and Methods for Testing, In: A. Steinbuchel (Ed.), CPL Press, UK, 2003, p. 366 (Chapter 12).

[3] Solanyl Biopolymer Inc., Canada, <http://solanylbiopolymers.com/what-is-biodegradability.asp>.
[4] What is Biodegradability? WiseGeek, <http://www.wisegeek.com/what-is-biodegradability.htm#>.
[5] L. Avérous, E. Pollet, Biodegradable polymers, environmental silicate nano-biocomposites, Green Energy and Technology, Springer-Verlag, 2012 (Chapter 2).
[6] E. Chiellini, R. Solaro, Biodegradable Polymers and Plastics, Springer, New York, 2003.
[7] J.E. Mark, Physical Properties of Polymers Handbook, second ed., Springer, New York, 2007.
[8] S. Ebnesajjad, Handbook of Biopolymers and Biodegradable Plastics, Elsevier, Waltham, MA, 2013 (Chapter 11).
[9] New Applications of Biodegradable Plastics in Medical Sector Gain Acceptance, <http://www.plastemart.com/upload/Literature/New-applications-biodegradable-plastics-medical-sector-Hernia-repair-drug-delivery-medical-implants.asp>.
[10] N. McDonald, Standards, regulations and testing of biodegradable polymers, GPEC Conference, 2013<http://sperecycling.org/sites/sperecycling.org/files/papers/GPEC2013/gpec2013-s5-mcdonald.pdf>.
[11] U. Pagga, Testing biodegradability with standardized methods, Chemosphere 35 (12) (1997) 2963–2972.
[12] Biodegradability Testing, Respirtek Consulting Laboratory, <http://www.respirtek.com/services/prod.htm>.
[13] Measuring Biodegradability, Science Learning, <http://sciencelearn.org.nz/Contexts/Enviro-imprints/Looking-Closer/Measuring-biodegradability>.
[14] E. Rudnik, Biodegradability Testing of Compostable Polymer Materials, Elsevier, 2008 (Chapter 11).
[15] R. Narayan, Fundamental principles and concepts of biodegradability—sorting through the facts, hypes, and claims of biodegradable plastics in the marketplace, Bioplastics Magazine 4 (January 2009), Germany. <www.bioplasticsmagazine.com>.
[16] ASTM Standard D 6400-99, Standard Specification for Compostable Plastics, West Conshohocken, PA, 1999.
[17] ASTM Standard D 5338, Standard Test Method for Determining Aerobic Biodegradation of Plastic Materials Under Controlled Composting Conditions, West Conshohocken, PA, 1998.
[18] International Standard ISO 14855, Determination of the Ultimate Aerobic Biodegradability and Disintegration of Plastic Materials Under Controlled Composting Conditions—Method by Analysis of Evolved Carbon Dioxide, Geneva, Switzerland, 1999.
[19] G.F. Moore, S.M. Saunders, Advances in biodegradable polymers, Rapra Review Report 9 (2) (1997) 9 (iSmithers Rapra Publishing).

[20] T. Kijchavengkul, R. Auras, M. Rubino, M. Ngouajio, R.T. Fernandez, Development of an automatic laboratory-scale respirometric system to measure polymer biodegradability, Polym. Test. 25 (2006) 1006–1016.
[21] G. Kale, Biodegradation of commercially available biodegradable packages in real and simulated composting Conditions (Master of Science Thesis), School of Packaging, Michigan State University, 2007.
[22] Novel Device for Aerobic Biodegradability Testing of Natural and/or Synthetic Polymers, Copolymers or Mixtures in an Aqueous Medium, Wetlands Biosciences S.A., Belgium, <http://www.umic-science.com/en/info/files/biodegradability-testing.php>.
[23] C. Bastioli, Handbook of Biodegradable Polymers, iSmithers Rapra Publishing, 2005.
[24] F. Wang, F.S.J. Lee, Poly(3-hydroxybutyrate) production with high productivity and high polymer content by a fed-batch culture of *Alcaligenes latus* under nitrogen limitation, Appl. Environ. Microbiol. 63 (1997) 3703–3706.
[25] Y. Tokiwa, C.U. Ugwu, Biotechnological production of (*R*)-3-hydroxybutyric acid monomer, J. Biotechnol. 132 (2007) 264–272.
[26] Y. Tokiwa, B.P. Calabia, Biological production of functional chemicals from renewable resources, Can. J. Chem. 86 (2008) 548–555.
[27] Y. Tokiwa, T. Ando, T. Suzuki, Degradation of polycaprolactone by a fungus, J. Ferm. Technol. 54 (1976) 603–608.
[28] Y. Tokiwa, T. Suzuki, Purification of polyethylene adipate-degrading enzyme produced by *Penicillium* sp. strain 14–3, Agric. Biol. Chem. 41 (1977) 265–274.
[29] Y. Tokiwa, T. Suzuki, Hydrolysis of polyesters by lipases, Nature 270 (1977) 76–78.
[30] Y. Tokiwa, Biodegradation of polycarbonates, in: S. Matsumura, A. Steinbuchel (Eds.), Miscellaneous Biopolymers and Biodegradation of Polymers, vol. 9, WILEH-VCH Verlag GmbH, Weinheim, Germany, 2002.
[31] Y. Tokiwa, T. Suzuki, Hydrolysis of copolyesters containing aromatic and aliphatic ester blocks by lipase, J. Appl. Polym. Sci. 26 (1981) 441–448.
[32] Y. Tokiwa, T. Suzuki, T. Ando, Synthesis of copolyamide-esters and some aspects involved in their hydrolysis by lipase, J. Appl. Polym. Sci. 24 (1979) 1701–1711.
[33] Y. Tokiwa, B.P. Calabia, C.U. Ugwu, S. Aiba, Biodegradability of plastics, Int. J. Mol. Sci. 10 (2009) 3722–3742, <http://www.ncbi.nlm.nih.gov/pmc/articles/PMC2769161/pdf/ijms-10-03722.pdf>.

5 Types of Biodegradable Polymers

5.1 Introduction

Rise of environmental pollution by synthetic polymers in developing countries have reached dangerous levels. Plastics produced from petroleum resources are not biodegradable. Because they defy microbial degradation, they end up in the landfills and damage the environment. To top that off, oil prices have increased remarkably. These are the two primary reasons which have brought interest in biodegradable polymers. Although, biodegradable plastics and polymers were first introduced in 1980s, they recently started to get traction, predominantly due to two major reasons: firstly environmental concerns, and secondly the realization that our petroleum resources are finite. Biodegradable polymers are broadly classified into two main categories: (1) natural and (2) synthetic. Natural polymers are available in large quantities from renewable sources, while synthetic polymers are produced from nonrenewable petroleum resources. Biodegradable polymers have a shelf life and involve biodegradation process. Biodegradation of polymers happen mostly due to cleavage of bonds through hydrolysis or by enzyme action in the polymer leading polymer erosion. There are many biodegradable polymers that have been produced along with microorganisms and enzymes capable of degrading them have been identified [1].

Biodegradable polymers, such as starch and cellulosic have been around for thousands of years and were typically used for food, furniture, and clothing. It was only in the last two decades, more and more products have been commercialized. Manufacturers are going green by modifying and improving their processes. For example, production of green polyethylene (PE), polyvinyl chloride, and other green plastics. Recent, technological advancements have substantially improved the properties of some bio-based polymers, such as heat resistant polylactic acid (PLA), enabling a wider range of applications. In addition, plants are being optimized, especially to provide biofibers with more stable resource properties over time. Wide range of applications have emerged recently where biodegradable polymers are sustainable options. Some of the applications include: packaging, biomedical products, textiles, agriculture, and household use [2].

Introduction to Bioplastics Engineering.
DOI: http://dx.doi.org/10.1016/B978-0-323-39396-6.00005-1

Biodegradable polymers represent a growing field. A vast number of biodegradable polymers (eg, cellulose, chitin, starch, polyhydroxyalkanoates (PHAs), polylactide, Poly(ε-caprolactone), collagen, and other polypeptides) have been synthesized or are formed in natural environment during the growth cycles of organisms. Some microorganisms and enzymes capable of degrading such polymers have been identified.

Biodegradable polymers are classified according to their origin into three classes: naturally produced renewable polymers, synthetic polymers derived from renewable resources, and synthetic polymers derived from petroleum-based resources. Some of the commercially produced biodegradable polymers are shown in Table 5.1 [3].

Natural biodegradable polymers are produced in nature by all living organisms. Biodegradation reactions are typically enzyme-catalyzed and occur in aqueous media. Natural macromolecules containing hydrolyzable linkages, such as protein, cellulose, and starch are generally susceptible to biodegradation by the hydrolytic enzymes of microorganisms. Cellulose, lignin, starch, chitin, collagen, gelatin, hyaluronic acid, dextran, heparin, xanthan, elastin, fibrin, pectin, and polyhydroxyalkanoates are biodegradable polymers naturally occurring in plants or animals.

Biodegradable polymers are primarily classified under two main groups as shown in Fig. 5.1 [4]:

A. Agro-polymers—As obtained from natural origin:
   - Polysaccharides (eg, starch, cellulose, lignin, and chitin)
   - Proteins (eg, gelatin, casein, wheat gluten, silk, and wool)
   - Lipids (eg, plant oils including castor oil and animal fats)
   - Polyesters produced by microorganisms or by plants (eg, polyhydroxy-alcanoates, poly-3-hydroxybutyrate)
   - Polyesters synthesized from bio-derived monomers (polylactic acid)
   - Miscellaneous polymers (natural rubbers, composites)
B. Biopolyesters—As obtained from mineral origin:
   - Aliphatic polyesters (eg, polyglycolic acid (PGA), polybutylene succinate (PBS), and Poly(ε-caprolactone)
   - Aromatic polyesters or blends of the two types (eg, polybutylene succinate terephthalate)
   - Polyvinylalcohols
   - Modified polyolefins (polyethylene or polypropylene with specific agents sensitive to temperature or light).

Table 5.1 Overview of Major Biodegradable Polymers and Their Suppliers [3]

Polymers	Polymers/trade name	Company
Poly(lactic acid)	NatureWorks PLA	Cargill, USA
	Lacty	Schimadzu, Japan
	PLA	Galactic, Belgium
	L-PLA	Purac, Netherlands
	PLA-based resins: Bio-Flex and Biograde	FKuR, Germany
	PLA	Cereplast
Polyhydroxyalkanoate (PHA)	Biomer	Biomer, Germany
	Nodax	Procter and Gamble, USA
	Mirel	Metabolix, USA
	ENMAT	TianAn Biologic Materials, China
Poly(ε-caprolactone) (PCL)	CAPA	Perstorp, UK
	Tone	Dow Chemicals, USA
	Celgreen	Daicel, Japan
Polyesteramide (PEA)	BAK	Bayer, Germany
Aliphatic polyesters	PBS	Anqing Hexing Chemical Co., China
	PBS	BASF, Germany
	PBS	Mitsubishi Gas Chemical, Japan
	PBS	Ire Chemicals, Korea
	PBS	Showa Highpolymer, Japan
Aliphatic copolyesters	PBSA Bionolle	Showa Highpolymer, Japan
	EnPol, PBSA	Ire Chemicals, Korea
	PBSA	Kingfa, China
	PBSA	IPC-CAS, China
Aromatic copolyesters (PBAT)	Ecoflex	BASF, Germany
	Biomax	DuPont, USA
	MATER-BI	Novamont, Italy
	Eastar Bio	Eastman Chemicals, USA
Starch	Solanyl	Rodenburg, Netherlands

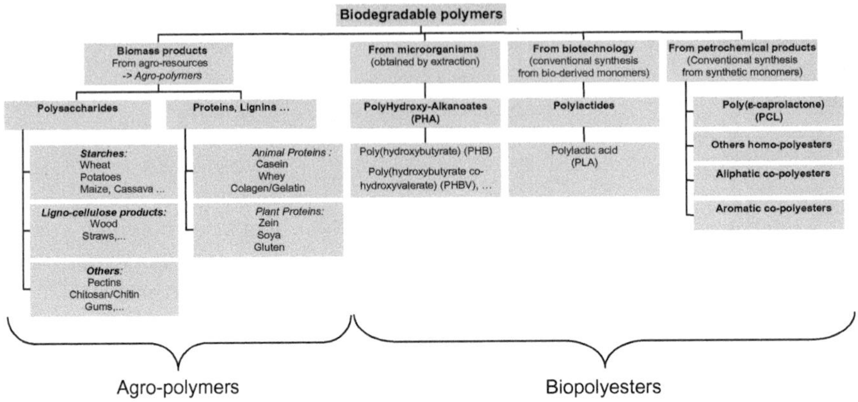

Figure 5.1 Classification of the main biodegradable polymers [4].

Table 5.2 Evolvement of Bio-Based Polymers [4,5]

Polymer generation	Updates
First generation	Focused on deriving polymers from agricultural feedstocks such as corn, potatoes, and other carbohydrate feedstocks
Second generation	Moved away from food-based resources

5.2 Bio-Based Polymers

Bio-based polymers are materials which are produced from renewable resources. Bio-based polymers can be biodegradable such as PLA or nondegradable such as bio-derived polyethylene (Bio-PE), also known as biopolyethylene. Similarly, while many bio-based polymers are biodegradable including starch and PHAs, there are some biodegradable polymers that are not bio-based such as Poly(ε-caprolactone) (PCL). Bio-based polymers offer important contributions. By reducing dependence on fossil fuels as well as reducing carbon foot print, bio-based polymers have done an important contribution [5].

Table 5.2 shows how bio-based polymers evolved. The first generation of bio-based polymers was derived mostly from agricultural feedstocks such as corn, potatoes, and other carbohydrates. However, because of advancements in biotechnology, second generation of bio-based polymers moved away from food-based resources. In essence, bio-based polymers similar to biodegradable polymers are produced by bacterial fermentation processes by synthesizing the building blocks (monomers)

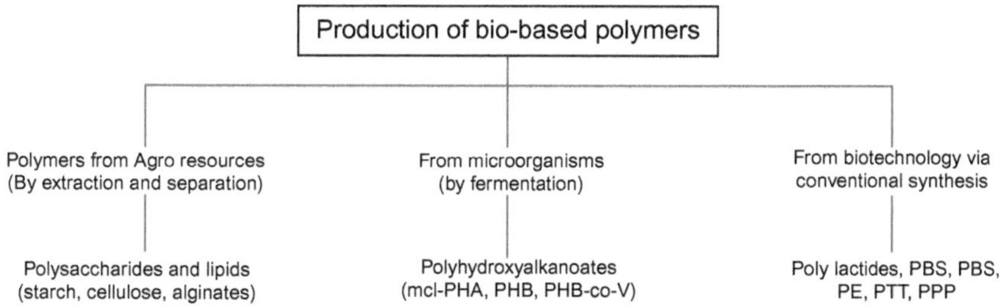

Figure 5.2 Different ways to produce bio-based polymers [4,5].

from renewable resources, including lignocellulosic biomass such as starch and cellulose, fatty acids, and organic waste. Natural bio-based polymers are the other class of bio-based polymers which are found naturally such as proteins, nucleic acids, and polysaccharides like collagen and chitosan [4].

There are three principal ways to produce bio-based polymers using renewable resources as shown in Fig. 5.2 [4,5]:

1. Using natural bio-based polymers with partial modification to meet the requirements (eg, starch)
2. Producing bio-based monomers by fermentation/conventional chemistry followed by polymerization (eg, PLA, PBS, and PE)
3. Producing bio-based polymers directly by bacteria (eg, PHAs).

Currently, there are start-up firms who have envisioned bringing bio-based polymers into commercial market. Invista and Genomatica both are pursuing to make polyamides intermediates from sugar. BASF, Cargill, and Novozymes are collaborating to make bio-based acrylic acid for superabsorbent polymers. And Coca-Cola achieved their goals when they came out with a 100% bio-based soda bottle [6]. Recently, Invista introduced a new version of its Lycra brand bio-based spandex. The spandex is 70% made from dextrose derived from corn. BASF has commercialized bio-based butadiene (bio-BDO) that can effectively be used in synthetic rubber. BASF has also engineered microbes to produce caprolactam, used in Nylon 6, and adipic acid and hexamethylenediamine, used in Nylon 6, 6 [6].

Bio-based polymers are broadly divided into two main categories:

a. Starch-based polymers
b. Cellulose-based polymers.

5.2.1 Starch-Based Polymers

Starch is primarily made up of two polysaccharides [7]: amylose, a mostly linear α-D (1,4′)-glucan and branched amylopectin, having the same backbone structure as amylose but with many α-1,6′-linked branch points as shown in Fig. 5.3 [8]. The starch chains has a lot hydroxyl groups, two secondary hydroxyl groups at C-2 and C-3 of each glucose residue, as well as one primary hydroxyl group at C-6 when it is not linked. The available hydroxyl groups on the starch chains can be oxidized and reduced, and can help in the formation of hydrogen bonds, ethers, and esters [9].

Starch comprises of 10–20% amylose and 80–90% amylopectin depending on the source [10]. Amylose is soluble in water, and forms a helical structure [11]. Short branched amylopectin chains are able to form helical structures which crystallize resulting in starch to exist as discrete granules. Starch granules exhibit hydrophilic properties and strong intermolecular association via hydrogen bonding formed by the hydroxyl groups on the granule surface. In amylose, α and β amylases can hydrolyze amylose; however, in amylopectin, α and β amylases can hydrolyze α-1→4 glycosdic linkage but cannot hydrolyze α-1→6 glycosdic linkage at branch points of amylopectin.

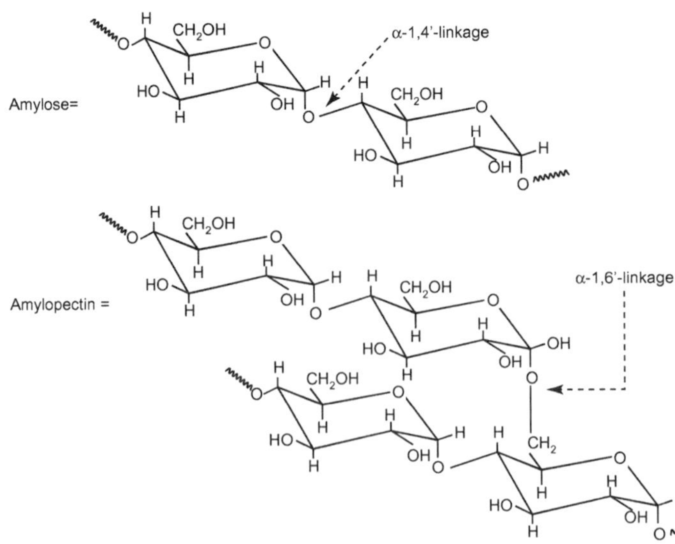

Figure 5.3 Molecular structure of starch [8].

Some of the market drivers of starch-based polymers are:

- lower cost materials than some other types of biodegradable polymers such as synthetic co-polyesters and PLA because of relatively cheap agricultural feedstock and simpler manufacturing process;
- environmental-friendly than synthetic biopolymers;
- starch blends have better physical and mechanical properties than pure plant based polymers.

Starch-based polymers are used in:

- applications which are used in natural environment such as agricultural and fishery materials;
- applications where reuse of the product is difficult and composting organic waste is effective;
- applications with specific features, where functionality and performance can also be completely separated from the main function.

Various types of bags and sacks such as refuse sacks, shopping bags, and compost bags are made from starch-based polymers. It has also found use in flexible packaging applications such as extruded bags and nets for fresh fruits and vegetables. Rigid packaging applications include thermoformed trays and containers for packaging fresh food and convenience food. Agriculture and horticulture applications include mulching film, covering film, and plant pots [12]. Fig. 5.4 shows different commercial applications made from starch-based polymers [13–18].

Starch-based polymers are typically classified into four types:

- Thermoplastic starch (TPS)
- Starch–synthetic aliphatic polyester blends
- Starch–PBS/PBSA polyester blends
- Starch–PVOH blends.

5.2.1.1 Thermoplastic Starch

TPS has gain attention and importance due to the fact that it has thermoplastic-like processability with temperature and shear; however the structures are more complex than those of synthetic thermoplastics [19,20].

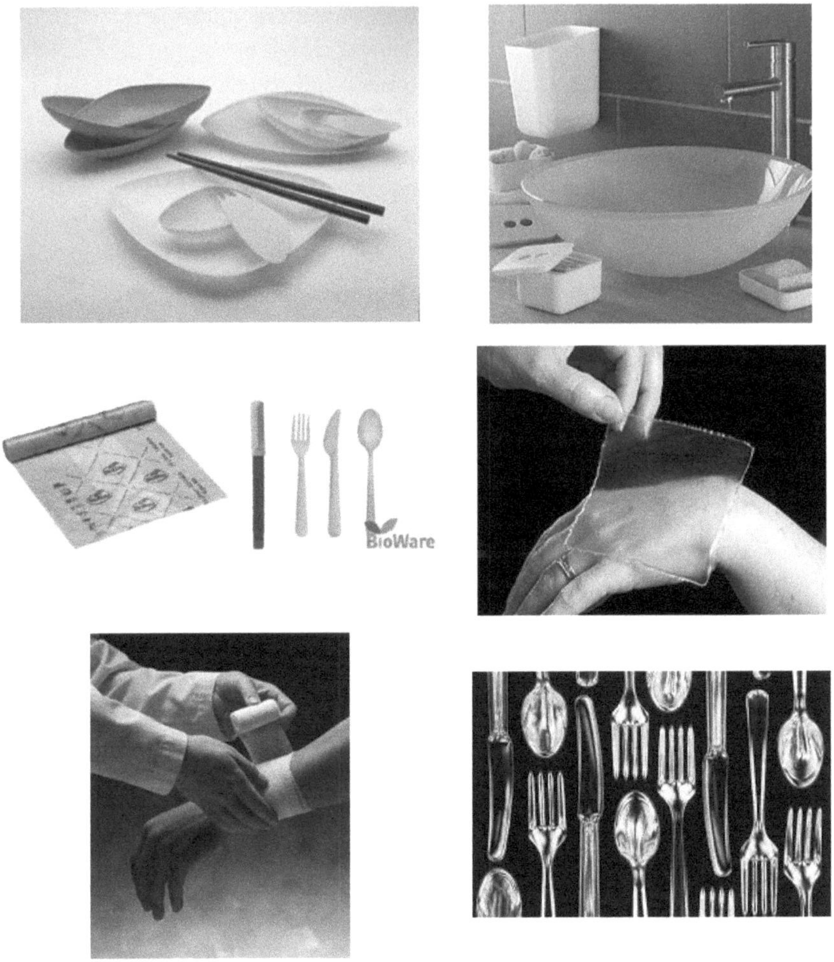

Figure 5.4 Commercial applications of products made from starch-based polymers [13–18].

TPS is similar to other polymers with linear and branched structures, molar mass, glass transition temperature, crystallinity, and melting temperature. However, in the presence of a plasticizer such as water, glycerin, sorbitol high temperatures (90–180°C) and shearing, it melts and fluidizes, enabling its use in injection, extrusion, and blowing equipment such as those for synthetic plastics [21,22].

By combining shear, temperature, and plasticization during extrusion, TPS melts similar to synthetic thermoplastic material [23]. Table 5.3 shows properties, initial and final formulation of wheat starch-based TPS when used at different plastication levels [23].

Table 5.3 Properties, Initial, and Final Formulation of Wheat Starch-Based TPS Using Different Plastication Levels [23]

% Starch	Moisture	Glycerol content (% starch)	Density	Glass transition temperature (°C)	Modulus (MPa)
74	16	14 (14)	1.39	43	997
70	12	26 (25)	1.37	8	52
67	9	36 (35)	1.35	−7	26
65	0	54 (50)	1.34	−20	2

By blending with other polymers, fillers, and fibers, properties of TPS can be dramatically improved. Some of the polymers used with TPS include:

- Cellulose
- Zein (a protein from corn)
- Natural rubber
- Polyvinyl alcohol (PVA)
- Acrylate copolymers
- Polyethylene and ethylene copolymers
- Polyesters
- Polyurethanes
- Poly(lactic acid)
- Poly(ε-caprolactone).

Blending of TPS with other polymers is often done using twin-screw extrusion at elevated temperatures. TPS when blended with other biodegradable polymers, such as PVA or aliphatic polyesters like PLA, PCL, and poly(3-hydroxybutyrate) (PHB) produce fully biodegradable polymer blend [4,20,23].

5.2.1.2 Starch Synthetic Aliphatic Polyester Blends

High-quality sheets and films for packaging are often made from blends of biodegradable synthetic aliphatic polyesters and starch. It is typical that approximately 50% of synthetic polyester is replaced with natural polymers, such as starch. Polyesters are also modified by incorporating different functional groups such as hydroxy, amine, and carbonyl that are capable of reacting with natural starch polymers [24].

Table 5.4 Commercially Available Starch-Based PCL Polymers [26]

Polymer type	Trade name	Supplier	Origin
Starch–Poly(ε-caprolactone) (PCL) blends	Mater-Bi Bioflex	Novamont Biotech	Italy Germany

Polycondensation of 1,4-butanediol and a mixture of adipic and succinic acids results in polyester. As an example, the wheat starch–polyester blends have a melting point close to the polyester alone. To make it flexible and more processable than the polyester itself, plasticizers are added to the starch. Plasticized blends were found to retain a high tensile strength and elongation at the break point, even at high concentration of starch [25].

When starch is blended with degradable polyesters such as PCL, the resulting blend is fully biodegradable. This has become focus of biodegradable polymer development. Typically, up to 45% of starch is blended with degradable PCL. Although, the blend is fully biodegradable, it is not strong enough for most applications. The melting temperature is relatively low around 60°C and it gets soft at temperatures above 40°C. Because of these drawbacks, starch–PCL has limited applications. Table 5.4 details some starch-based PCL polymers that are commercially available [26].

Even with limitation, there are several products made from starch-based plastics are currently on the market. These include films for shopping bags, bread bags, bait bags, overwraps, and backing materials for flushable sanitary products and mulch film.

Novamont resins Inc. introduced a product called BioBag which is a blend of corn starch and fully degradable PLA. This blended product, BioBag, is used in high-quality sheets and films for packaging and other film applications. There are other products that are produced from starch-based foams. One such example is water-soluble beads made from potato starch replacing polystyrene (PS) foam beads. Starch-based beads are readily soluble and biodegradable, have found a place in the market for biodegradable plastics [26].

Two of the major starch-based synthetic aliphatic polyester blends are:

- Starch–PBS/PBSA polyester blends
- Starch–PVOH blends.

5.2.1.3 Starch–PBS/PBSA Polyester Blends

One of the major starch-based synthetic aliphatic polyester blends are starch–PBS/PBSA polyester blends. PBS and polybutylene succinate adipate (PBSA) are synthesized from 1,4-butanediol and succinic and/or adipic acid at 215–225°C under high vacuum. The resulting average molecular weight of 40 kg/mol is not sufficient. In order to increase molecular weight to the desired level, a small amount of unsaturated carboxylic acid is added under addition polymerization initiated by peroxides [26].

Starch by itself has poor mechanical properties. PBS and PBSA polyester blends are mixed with starch to improve material mechanical properties. A small amount of compatibilizer is added to impart phase stability to the starch-based polyester blends. Typically, 5% by weight of compatibilizer is sufficient to provide stability to starch-based blends. Plasticizers are also added to starch-based blends to increase flexibility and reduce brittleness. PBSA is biodegradable, and exhibits excellent thermoplastic properties. Corn starch is an inexpensive polysaccharide that is blended with PBSA at concentrations of 5–30% by weight [27].

Starch–PBS/PBSA blends disintegrate in compost after 6 weeks. Some of their applications includes films for compostable trash bags, paper lamination, magnetic cards, sheets for thermoforming, extrusion-forming, monofilament for fishing lines, woven nets, and ropes [27].

5.2.1.4 Starch–PVOH Blends

Blending starch with biodegradable polyester results in phase separation and poor interfacial properties. Because PVOH degrades at high temperature when processed by melt processing, starch–PVOH blend uses solution casting to produce films. However, low efficiency and high processing cost makes solution casting economically not viable and hence not the process of choice.

Plasticizers like water and polyols are used to decrease the melting temperature of PVOH and increase the flexibility and workability of starch and PVOH. This makes starch–PVOH blend to be processed on thermal processes such as extrusion. In addition to water, glycerol and citric acid, there are other complex plasticizers that have been successfully used. Among the conventional plasticizers, glycerol is much more effective than water. However, glycerol has shown lower tensile strength and water barrier properties. In order to improve the properties, glycerol is mixed with water in 1:1 ratio. This is considered as the optimum plasticizer for starch–PVOH blends [28].

Plasticizer is the integral part of starch–PVOH film properties and can drastically affect by the type and the quantity of the plasticizer added. Typically, starch–PVOH film contains 51% starch, 17% PVOH, 30% plasticizer, 1% paraformaldehyde, and 1% NH_4Cl. When glycerol, sorbitol, and citric acid are used as plasticizers, they impact differently on the final properties of the product. Citric acid has proven to be a better plasticizer over glycerol. This is in part because citric acid produces hydrogen bonding with the hydroxyl group and carboxyl group and thus increased inter- and intramolecular interactions between starch, PVOH, and plasticizers [28].

Mechanical properties of starch–PVOH blends are directly impacted by the amount of the plasticizers added. Higher concentration of plasticizer drastically improves mechanical properties and reduces water absorbance. In the case of citric acid, mechanical properties such as tensile strength and elongation at break are improved drastically with increase in the citric acid concentration [20,29].

5.2.2 Cellulose-Based Polymers

Cellulose is a natural polymer made from long chains that are linked together by smaller molecules. These links in the cellulose chains consists of β-D-glucose. These sugar units are linked when water is eliminated by combining the H and hydroxyl group as shown in Fig. 5.5 [30].

When two of these sugars are linked together, they produce a disaccharide called as cellobiose. The structure of cellobiose is shown in Fig. 5.6 [31,32]. In this structure, the spatial arrangement of the acetal linkages is very important.

Fig. 5.7 shows two ring structures of the cellulose molecule. These structures are called the pyranose rings. In this ring structure, all the groups are larger than hydrogen which sticks out from the periphery of the rings [33]. The spatial arrangement of carbon atoms in the glucose molecule at positions 2, 3, 4, and 5 is fixed; however, in a ring structure such as pyranose, the hydroxyl group at C-4 can approach the carbonyl at C-1 from either side, thus resulting in two different spatial arrangements at carbon

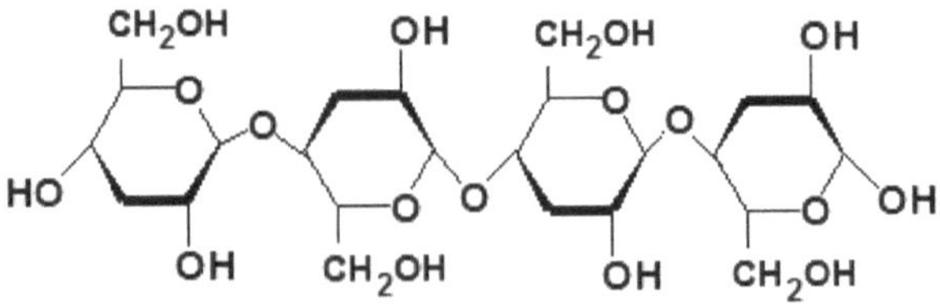

Figure 5.5 Structure of Cellulose [30].

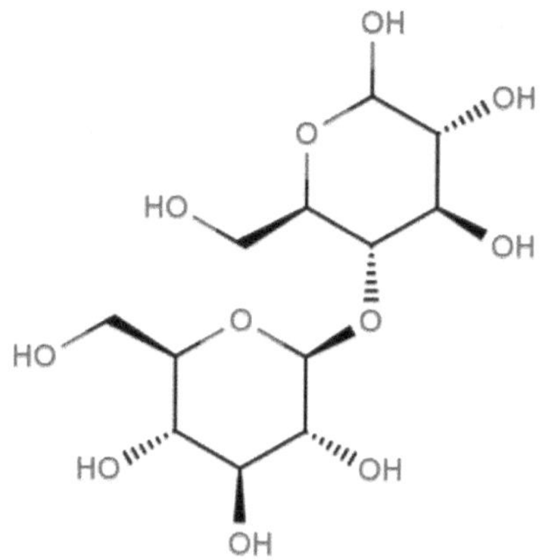

Figure 5.6 Structure of cellobiose [31,32].

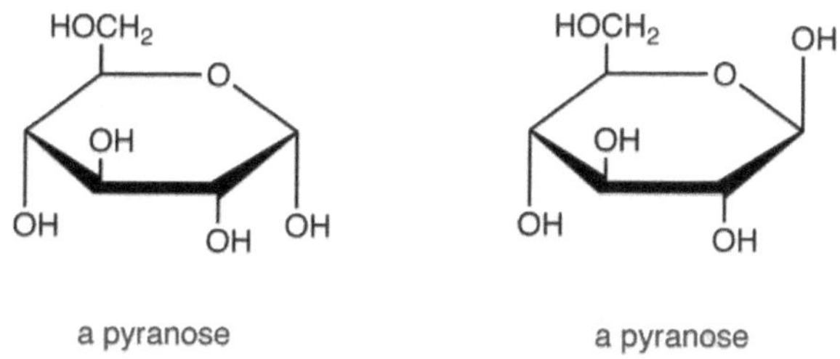

Figure 5.7 Structure of a pyranose, example, two of the four cyclic forms of D-glucose [33].

position 1. When the hydroxyl group at C-1 is on the same side of the ring as the C-6 carbon, it is said to be in α configuration. In cellulose, the C-1 oxygen is in the opposite or β configuration [34].

Because of the alignment of the hydroxyl groups on the cellulose chain, they tend to stick out laterally along the extended molecule and are readily available for hydrogen bonding. The chains are grouped together in a highly ordered structure by means of hydrogen bonding. Due to long length of the chains, they tend to pass through several crystalline regions with some areas of disorder in the middle. These disordered middle areas are termed as fringed-micelle regions [35,36]. Because of the strong hydrogen bonding in the crystalline regions, cellulose has good strength and resistance toward solvents. This also prevents them from melting. In comparison, less-ordered regions, more hydrogen is available to bond with other molecules such as water because the chains are located far apart from one another. Most cellulose structures can absorb large quantities of water and hence they are hygroscopic. This makes cellulose to swell but does not dissolve in water [34].

There are two major cellulose-based polymers that are commonly used:

1. Cellulose esters
2. Celluloid

5.2.2.1 Cellulose Esters

Cellulose esters are part of a large family of cellulose derivatives that have found use in pharmaceutical and other applications. Cellulose ester is divided into two categories—enteric and nonenteric. Enteric esters are those which are relatively insoluble in acid solutions but soluble in mildly acidic to slightly alkaline solutions such as cellulose acetate phthalate (CAP). However, nonenteric esters are not dependent on pH solubility characteristics. They are mostly insoluble in water with the exception of cellulose acetates (CAs) with low level of acetyl [37].

Acid catalysis of an acid and alcohol also known as Fischer esterification process is the most important method for preparing an ester. In comparison, traditional esterification method uses controlled-acid catalysis of an acid or acid anhydride and the hydroxyl groups of cellulose to make CA. Because this is a reversible reaction, the reaction requires complete removal of water for completion. Degree of substitution (DS) is determined by the amount of water. The general reaction to make cellulose ester is shown as illustrated reaction:

$$\text{Cellulose} + \text{acid} \xrightarrow{\text{catalyst}} \text{cellulose ester} + \text{water}$$

Cellulose esters are commercially available for a variety of applications. Some of the applications include solvent-borne coatings, separation, medical and controlled release, composites and laminates, and plastic applications.

The three most common types of cellulose esters are:

1. Cellulose acetate (CA)
2. Cellulose acetate propionate (CAP)
3. Cellulose acetate butyrate (CAB).

These three cellulose esters are thermoplastic materials that are produced through esterification of cellulose. In order to make powdered cellulose esters biopolymer, different raw materials such as cotton, recycled paper, wood cellulose, and sugarcane are needed. For example,

Table 5.5 Melting Points of Cellulose Esters [39]

Cellulose ester	Melting point (°C)
Cellulose	450
Acetate	306
Butyrate	183
Caproate	94
Caprylate	86
Caprate	88
Heptylate	88
Laurate	91
Myristate	106
Palmitate	106
Propionate	234
Valerate	122

Bioceta, plasticized CA, is prepared from cotton flakes and wood pulp through an esterification process with acetic anhydride. CAP and CAB are mixed esters produced by treating cellulose with appropriate acids and anhydrides in the presence of sulfuric acid [38]. Table 5.5 shows melting points of cellulose esters [39].

5.2.2.2 Cellulose Acetate (CA)

CA is the first organic ester of cellulosic family. It was originally discovered by Schutzenberger in 1865 and later developed by Franchimont in 1879 and Miles in 1903. By 1929, commercial grades of CA in the form of molding powder, fibers, sheets, and tubes were being produced in the United States [40].

Fig. 5.8 shows general mechanism to prepare CA [41,42]. CA is prepared by mixing cellulose with acetic anhydride using acetic acid as solvent and sulfuric acid as catalyst. Sulfuric acid reacts with acetic anhydride to form acetylsulfuric acid. During the acetylation process, both sulfuric acid and acetylsulfuric acid react with cellulose to form cellulose sulfate acid ester.

Most of the sulfate groups are replaced by the acetyl groups during the acetylation process. This mixture is allowed to react until the three hydroxyl groups per anhydroglucose have been replaced with the acetyl groups. In the next step, hydrolysis is performed to remove acetyl and sulfate groups under controlled conditions of time, temperature, and acidity.

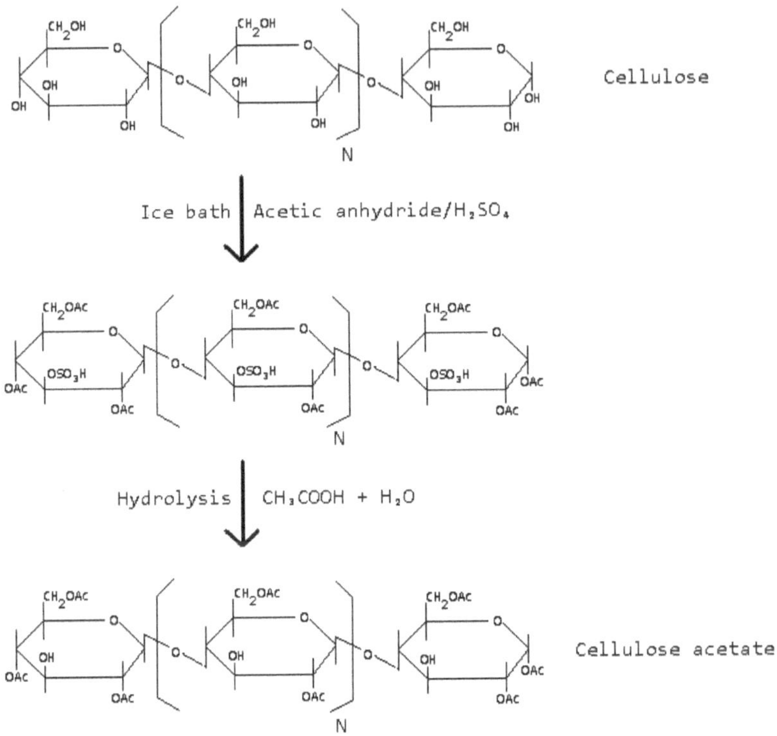

Figure 5.8 General mechanism to prepare cellulose acetate [41,42].

This helps in improving the physical properties of CA. Hydrolysis process also removes the combined acid sulfate ester and thereby improving the thermal stability of CA [41,42].

A pretreatment of cellulose with water or acetic acid is performed to make more rapid and uniform ester. This pretreatment swells the fiber and thereby reducing intramolecular hydrogen bonding. The rate of acetylation of preswollen cellulose is about three times higher than that of nonswollen cellulose. It also makes them accessible to the esterifying mixture, and hence increases their reactivity.

Table 5.6 shows some properties of CA. CA is fairly inexpensive material and is either in transparent or colored. They exhibit poor thermal, electrical, and weathering properties, and are not resistant to chemicals. They are typically used in films and sheets in packaging and display industries. CAs are easily processed and have been used in brush handles, combs, and spectacle frames. They are also used in vacuum-formed

Table 5.6 Properties of Cellulose Acetate [43]

Property	Value
Molding qualities	Excellent
Relative density	1.22–1.34
Tensile strength (MPa)	13–62
Compressive strength (MPa)	14–248
Impact strength, Izod (J/mm)	0.02–0.26
Hardness, Rockwell	R34–R125
Thermal expansion (10^{-4}/°C)	20–46
Resistance to heat (°C)	60–105
Dielectric strength (V/mm)	9840–23,620
Dielectric constant (at 60 Hz)	3.5–7.5
Dissipation factor (at 60 Hz)	0.01–0.06
Arc resistance (s)	50–310
Water absorption, 24 h (%)	1.7–6.5
Burning rate (mm/min)	Slow to self-extinguishing
Effect of sunlight	Slight
Effect of acids	Decomposes
Effect of alkalis	Decomposes
Effect of solvents	Soluble in ketones, esters, chlorinated hydrocarbons, and aromatic hydrocarbons
Machining qualities	Excellent
Optical qualities	Transparent to opaque

display containers for hardware or food products, and barrier films. CA coated films are used in magnetic recording tapes and photographic film. CA plastics are made into fibers for use in textiles [43].

The DS of acetyl groups per anhydro-D-glucose unit of cellulose can be calculated from the acetyl content as follows [43]:

$$DS = \frac{3.83 \times \%\ \text{acetyl}}{102.4 - \%\ \text{acetyl}}$$

Because the formula weight is 60 for CH_3COOH and 43 for CH_3CO, the ratio between these two is 1.395. Hence, the relationship between acetyl content and combined acetic acid can be obtained by using the following equation:

$$\text{Combined acetic acid} = \text{acetyl content} \times 1.395$$

Table 5.7 Relationship of Cellulose Acetate DS to Acetyl Content and Combined Acetic Acid [43]

DS	Acetyl content	Combined acetic acid
0.50	11.7	16.3
0.75	16.7	23.3
1.0	21.1	29.4
1.5	28.7	40.0
2.0	35.0	48.8
2.5	40.3	56.2
3.0	44.8	62.5

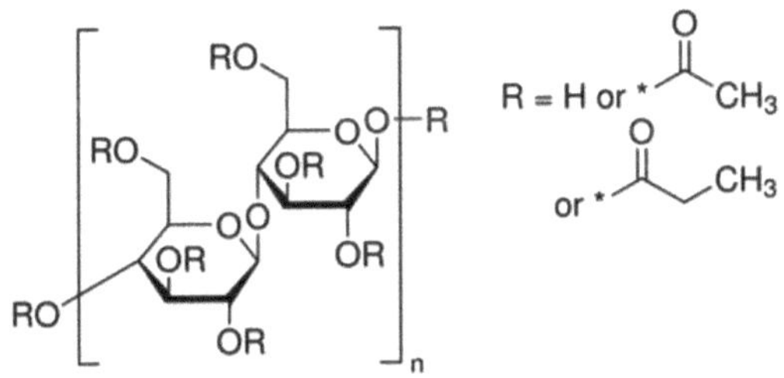

Figure 5.9 General structure of cellulose acetate propionate [44].

Table 5.7 shows relationship of CA DS to acetyl content and combined acetic acid. The thermoplastic characteristics of CA are greatly improved as the acetyl content is increased from around 20% (DS = 1) to around 39% (DS = 2.4). Higher acetyl content gives poorer flow but better moisture resistance; lower acetyl content gives better impact strength. The melting point of CA also drops from 300°C to approximately to 230°C with decrease in acetyl content. The high melting point can potentially lead to thermal degradation of cellulose ester. In order to prevent this degradation during processing, plasticizers are added which helps in lowering the melting point [43].

5.2.2.3 *Cellulose-Acetate Propionate (CAP)*

CAP was originally developed by the Celanese Plastics Company in 1931. Similar to other acetates, it is made with the addition of propionic acid (CH_3CH_2COOH) in place of acetic anhydride. The general structure is shown in Fig. 5.9 [44]. Plastic grade CAP has an acetyl content in the range of 1.5–7 wt% and a propionyl content of about 39–42 wt%.

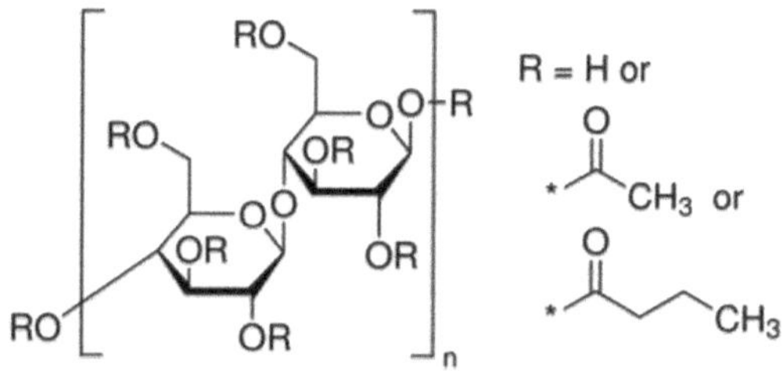

Figure 5.10 General structure of cellulose acetate butyrate (CAB) [45].

Because of its high transparency, low-level light scattering, and good impact resistance, CAP is used in high-quality frames for sun glasses, personal protective equipment, and sport goggles.

5.2.2.4 *Cellulose-Acetate Butyrate*

Hercules Powder Company and Eastman Chemical jointly developed CAB during the mid-1930s. Fig. 5.10 shows a general structure of CAB [45]. CAB is produced when cellulose is reacted with a mixture of sulfuric and acetic acids followed by esterification process. During the esterification process cellulose is reacted with butyric acid and acetic anhydride. This reaction is similar to the ones used in producing CA except that butyrate is also used. The end product has acetyl groups (CH_3CO) and butyl groups ($CH_3CH_2CH_2CH$) in the repeating cellulose unit.

CAB has acetyl content in the range of 13–15 wt%, a butyryl content of 34–39 wt%, and a free hydroxyl group of 1–2 wt%. CAB products have good dimensional stability among family of other acetates. They have excellent toughness, moisture resistance, and are available in fine colors. They are typical extruded and have found their use for automobile steering wheels, knobs, tool handles, packaging blisters, illuminated advertising signs, machine hoods, lamp covers, and dome lights.

5.2.2.5 *Celluloid*

In the middle of the 19th century during the times of early experimentation and synthesis, Celluloid was invented. It all started when cellulose was treated with strong nitric acid to form nitrocellulose, which found some use as an explosive. During 1860 to 1870, English chemists tried to stabilize the process with different solvents. They were somewhat successful, however, the process proved to be unprofitable. Later, an American scientist developed a material which was produced by mixing camphor with nitrocellulose. The camphor dissolved the nitrocellulose,

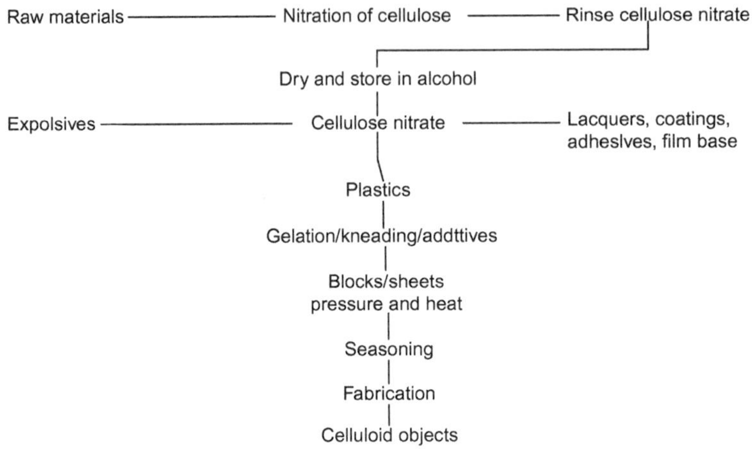

Figure 5.11 Flow chart of celluloid manufacture [47,48].

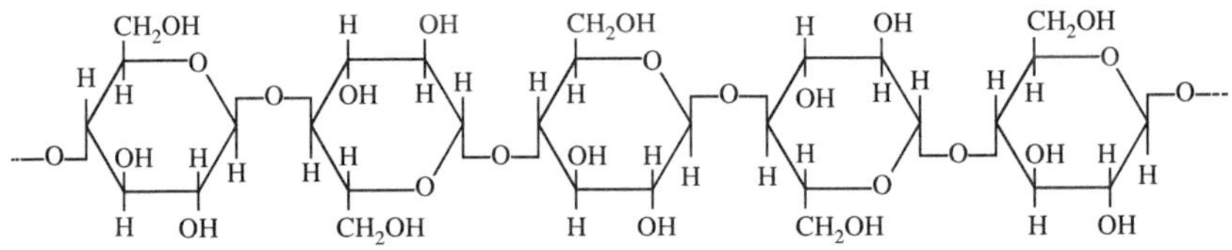

Figure 5.12 Structure of cellulose [47,48].

but also affected the structure of the product and performed as a plasticizer. This development was ultimately successful and became trademarked as Celluloid [46].

Fig. 5.11 shows the flow chart of celluloid manufacture. The raw materials required to synthesize celluloid include cellulose fibers, nitric acid, sulfuric acid, water, alcohol, camphor, colorants, and fillers [47,48].

Cellulose, the starting raw material is obtained from a variety of sources, including cotton, ramie, flax, jute, and wood pulp. The structure of cellulose is shown in Fig. 5.12 [47,48]. In the early days, the main commercial source of cellulose for cellulose nitrate synthesis in the United States was cotton mill waste, including cloth scraps from clothing production and waste fibers from spinning plants.

When a cellulose molecule (Fig. 5.13) is chemically modified by strong nitric acid, sulfuric acid, and water, the resultant product is cellulose nitrate. The structure of cellulose nitrate is shown in Fig. 5.13 [47,48]. Cellulose nitrate has been found useful in a variety of materials. Some of them include adhesives, explosives, film bases, lacquers

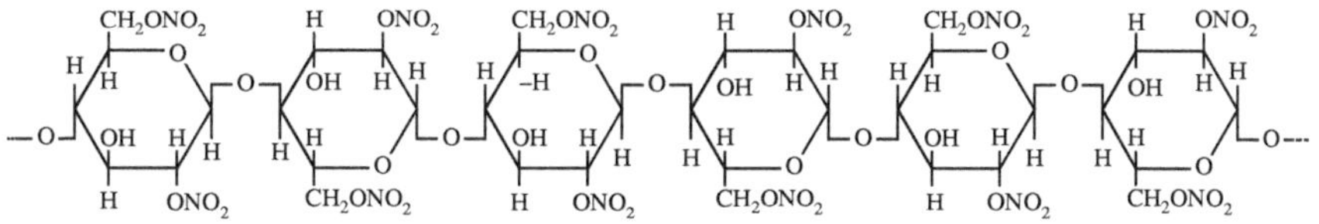

Figure 5.13 Structure of cellulose nitrate [47,48].

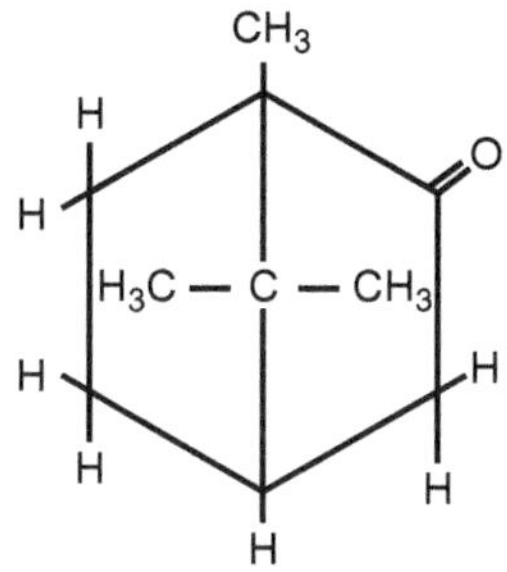

Figure 5.14 Structure of camphor [47–49].

and coatings, and solid plastics. Depending on the degree of nitration of the cellulose nitrate molecule used in manufacture can result in different products. In addition, by varying the ratio of cellulose nitrate to plasticizer and the varying amounts of solvent in the formulation can also lead to a different product. The level of nitration of a cellulose nitrate molecule is usually given as the percentage of nitrogen atoms per glucose residue in the molecule [47,48].

Once cellulose nitrate is synthesized, it is then mixed with camphor; a resin from the evergreen *Cinnamomum camphora*. The structure of camphor is shown in Fig. 5.14 [47,49]. Camphor is a bicyclic material belonging to terpenoid family. It is white and crystalline substance with a specific gravity of 0.986 to 0.996. Camphor sublimes at room temperature, melts at 176°C and boils at 209°C [48,49]. Synthetic camphor is produced by converting α-pinene from turpentine to camphene to isobornyl acetate to isoborneol to camphor. Synthetic camphor is chemically identical to natural camphor but is optically inactive (racemic) [50].

Some of the characteristics that make camphor the best solvent for cellulose nitrate are as follows [47–49]:

- Excellent gelatinizing agent of quick and deep penetration.
- Readily soluble in ethyl alcohol, methyl alcohol, and acetone.

- It has a high melting point (176°C) and boiling point (209°C).
- Unpleasant odor but no cumulative or deleterious effects on the makers and handlers of the celluloid.
- Volatility is extremely low.
- Solvent resistant to harsh chemical such as nitric acid, acids, and alkalis.
- Reduces the explosiveness of cellulose nitrate when mixed with it.
- Does not separate out of celluloid upon heating and cooling or upon long storage.
- Light has no effect upon it.

Colorants such as pigments and dyes can be effectively added to celluloid without much change in its properties. Some other fillers and additives include alcohol-soluble antacids for translucent celluloid. Zinc oxide and zinc carbonate are added to opaque celluloid to make it stabilized and do not require added antacid [50]. Stabilizers for translucent celluloid included aryl phenyl ureas, lactates (sodium, potassium, calcium, strontium, magnesium), and many other substances [51].

Fig. 5.15 shows thermal degradation process of celluloid [47]. Typical degradation process involves recrystallization of cellulose nitrate molecules and the expulsion and sublimation of camphor. This results in product getting brittle and noticeable crazing in transparent or translucent products. In extreme cases such as degradation of nitrocellulose, it gives off oxides of nitrogen which in contact with water form nitric acid. Nitric acid is very corrosive to metal, fabric, paper, etc., and often results in substantial destruction of surrounding materials. The effects are worse when the celluloid products are stored in closed containers. The fumes are more concentrated, and the by-product catalyzes further reaction, leading to further damage [47].

In order to prevent degradation of celluloid, proper storage of celluloid in ventilated containers without excess humidity or high temperatures is required. However, once the degradation process has started, it is impossible to reverse. Coloring agents that prevent light transmission through celluloid objects appear to give some protection.

$$\begin{aligned} RCH_2CH_2ONO_2 + H_2O &\rightarrow RCH_2CH_2OH + HNO_3 \\ + OH^- &\rightarrow RCHCH_2 + H2O + NO_3^- \\ &\rightarrow RCH2CHO + HNO_3 \end{aligned}$$

Figure 5.15 Thermal degradation of celluloid [47].

Some of the factors that can contribute to degradation of celluloid include [47]:

- the purity of ingredient materials;
- the rinsing and drying processes, which may leave agents catalyzing degradation in the finished materials;
- the kneading process.

5.2.3 Chitin and Chitosan

Chitin is the second most abundant biodegradable polymer produced in nature after cellulose. It is an acetylated polysaccharide composed of *N*-acetyl-D-glucosamine groups linked by β (1→4) linkages and exists as ordered crystalline microfibrils shown in Fig. 5.16a [52].

Deacetylation of chitin results in chitosan. Chemical structure of chitosan is shown in Fig. 5.16b [52]. Chitin is insoluble in its native form but chitosan, is water soluble. Enzymes, such as chitosanase, lysozyme, and papain are known to degrade chitosan. The rate of degradation of chitosan inversely depends on the degree of acetylation and crystallinity of the polymer [53].

Chitosan and chitin are biocompatible and have antimicrobial properties and ability to absorb heavy metal ions. Because of their water-retaining and moisturizing properties, they are used in cosmetic industry [54].

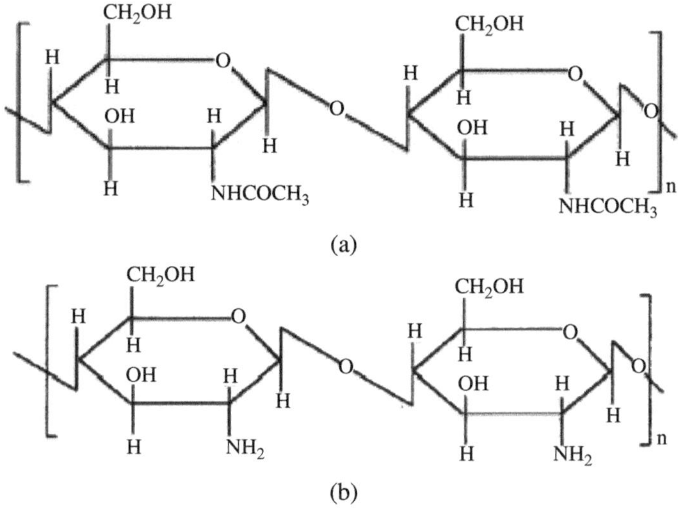

Figure 5.16 Chemical structure of (a) chitin and (b) chitosan [52].

Chitosan can be formed into membranes and are used in tissue-engineering applications [55]. The high chemical reactivity of chitosan makes them useful in drug conjugates for cancer therapy [56]. Chitosan gels, powders, films, and fibers have found use in encapsulation, membrane barriers, contact lens materials, and cell cultures [57].

5.2.4 Bacterial Polyesters

Among all biodegradable polymers, polyesters are considered as a primary choice because of their hydrolyzable ester bonds. The polyesters are classified into two types: aliphatic polyesters and aromatic polyesters [58]. Some bacterial polyesters utilize bacteria like Bacillus megaterium, Wautersia eutropha, and Pseudomonas species to synthesize polyesters however, there are some polyesters that can be synthesized by bacterias such as metabolites. There are certain bacteria that can accumulate up to 80% of their dry weight before autohydrolysis.

Some of the important characteristics of bacterial polyesters are as follows [59]:

1. They are water-resistant, and products made of the polyesters are water-tight.
2. The material can be processed by injection and blow molding.
3. Polyesters are not flexible and tend to become brittle.
4. Polyesters tend to lose their vapor barrier properties.

Some of the bacterial polyesters are:

- Polyhydroxyalkanoate (PHA)
- Polyhydroxybutyrate (PHB)
- Poly(hydroxy- butyrate-hydroxyvalerate) (PHB/HV)
- Poly(ε-caprolactone) (PCL).

Table 5.8 shows the list of bacterial polyesters with their specified radical [60].

5.2.4.1 Polyhydroxyalkanoates

PHAs are a family of intracellular biopolymers produced by bacterial fermentation of sugar or lipids. They are produced by numerous bacteria to store carbon and energy. The structure of PHAs is shown in

Table 5.8 Radical of Bacterial Polyesters [60]

Polymer/grade (Company)	Radical
PHA	–
PHB	CH_3
PHV	$-CH_2-CH_3$
PHBHx (Kaneda)	$-CH_2-CH_2-CH_3$
PHBO (Nodax)	CH_3 or/and $-(CH_2)_4-CH_3$
PHBOd (Nodax)	CH_3 or/and $-(CH_2)_{14}-CH_3$

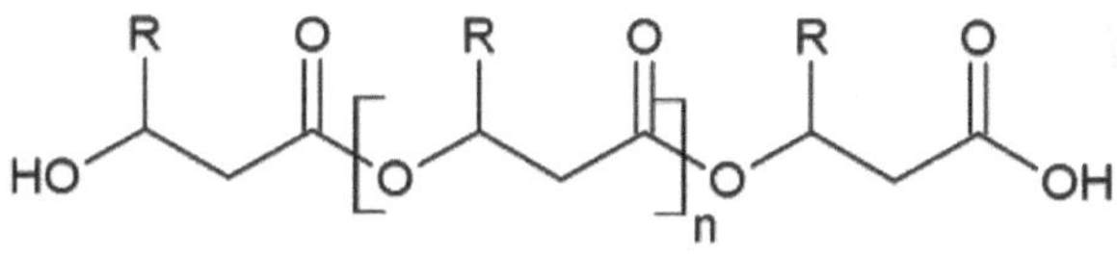

Figure 5.17 Structure of Poly(ε-caprolactone) (PHA) [61].

Fig. 5.17 [61]. Molecular weight of these polyesters can range from tens into the hundreds of thousands depending on growth conditions, bacterial strain, and carbon source [57]. Bacterially synthesized PHAs have found importance because they can be produced from a variety of renewable resources and are considered biodegradable.

PHAs are formed mainly from saturated and unsaturated hydroxy-alkanoic acids. The monomer of PHA can be branched or unbranched 3-hydroxyalkanoic acids or those with substituted side chains as 4- and 5-hydroxyalkanoic acids. PHAs are homo-, co- and terpolymers depending on the kind of monomer. Wide variety of PHAs with different properties can be created from variety of monomers and by varying constitutional isomerism.

Fig. 5.18a and b shows process of PHA synthesis from bacteria. Accumulation of PHA is an inherent response to the stress conditions faced by bacterial cells. Stress conditions faced by bacterial cells are generated in vitro by exposing bacteria to nutrient limitations, due to which they switch their metabolic pathways and cause PHA production as their carbon and energy reserves to name a few of these substrates are bagasse, molasses, corncob, and other agricultural wastes.

Different microorganisms can be used to produce PHAs. There are more than 300 different microorganisms that generate PHAs as natural energy reserves. PHAs tend to degrade if they lack carbon or energy. Microorganism's stability and biological safety, its PHA production

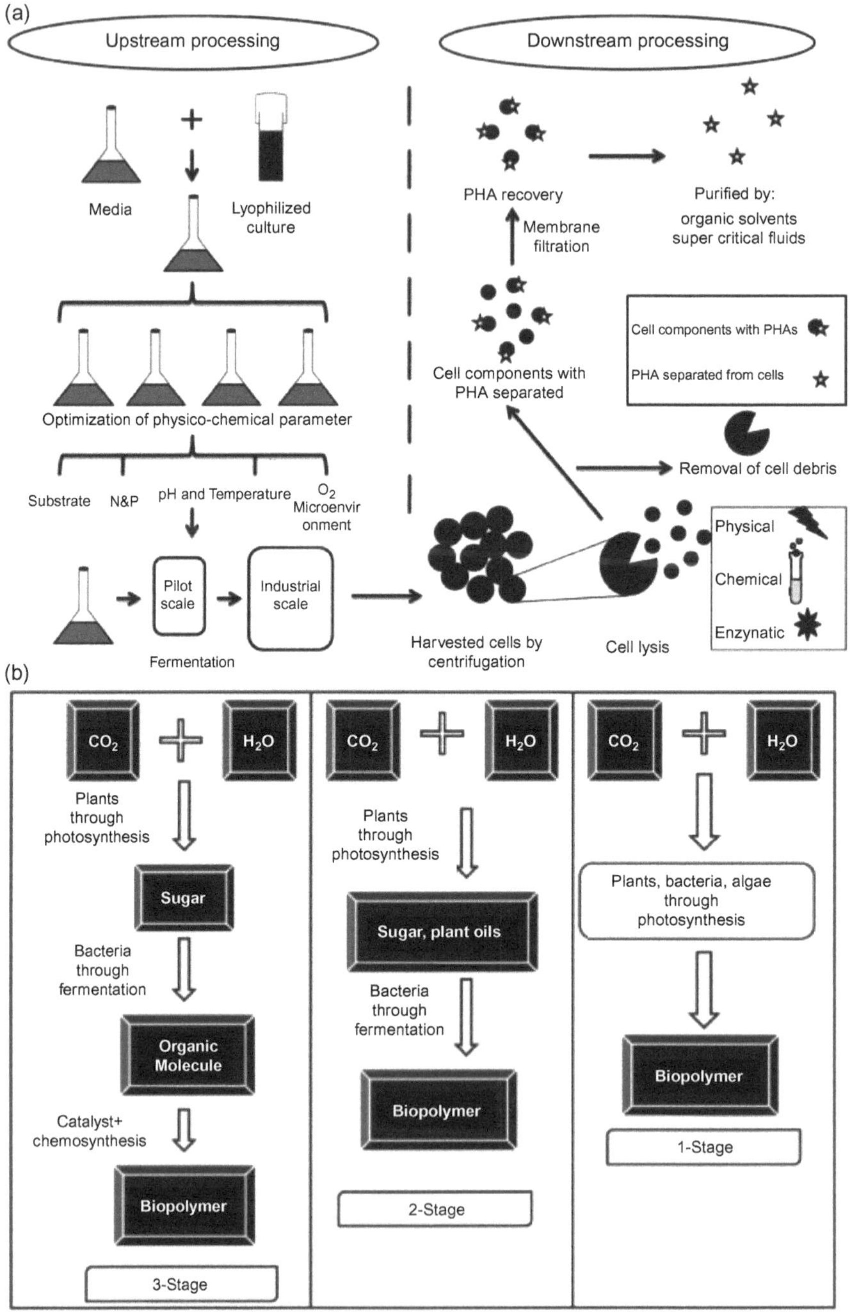

Figure 5.18 (a) Synthesis of PHA and (b) systems for PHA synthesis [62].

rates, PHA extractability, the molecular weight of the agglomerated PHA as well as the spectrum of useable carbon sources will dictate the type of microorganism for an application is needed [63].

The first step of the bacterial fermentation process is inoculation. In this process, the bacteria required for the subsequent metabolization process multiply and grow in an aqueous medium enriched with a balanced nutrition supply and air under optimum physical conditions. The next step involves actual PHA synthesis under conditions not conducive to growth and multiplication and a relative oversupply of carbon. The PHAs are usually stored in intracellular inclusion bodies and can account for up to 90% of dry cellular weight. Their molecular weight generally ranges from 100,000 to 500,000 g/mol. However, molecular weights of considerably more than 1,000,000 g/mol are obtained under special conditions. The complete fermentation process typically takes 2 days [63].

Some of the general characteristics of PHAs are as follows:

- Water insoluble and relatively resistant to hydrolytic degradation
- Good ultraviolet resistance but poor resistance to acids and bases
- Soluble in chloroform and other chlorinated hydrocarbons
- Biocompatible and hence suitable for medical applications
- Sinks in water, facilitating its anaerobic biodegradation in sediments
- Nontoxic
- Less sticky than traditional polymers when melted.

The average properties of PHA's are summarized in Table 5.9 [63].

PHAs are insoluble in water but still degradable and biocompatible. They can be processed well by injection molding. PHAs exhibit good barrier properties against oxygen and, compared to other biopolymers, a slightly higher barrier effect against water vapor. PHAs can be used in a variety of applications ranging from industrial to medical. Fig. 5.19 shows a bicycle helmet made from knitted cellulosic high-performance fibers that had been used to reinforce Biopol B (9% 3-HV) to optimize the strength and energy absorbance behavior [64–66].

Fig. 5.20 shows PHAs is used in bio-implants, biofuels, drugs and fine chemicals, industrial fermentation, food and feeds, and bioplastics [67]. PHAs are also used as packaging films mainly in shopping bags, containers and paper coatings, disposable items such as razors, utensils,

Table 5.9 **Typical Properties of PHA's [63]**

Property (units)	Values
Glass transition temperature, T_g (°C)	2
Melting temperature, T_m (°C)	160–175
Degree of crystallinity, X_{cr} (%)	40–60
Young's modulus, E (GPa)	1–2
Tensile strength, σ (MPa)	15–40
Elongation at break, $\mathcal{E}$ (%)	1–15
Water vapor transmission rate, WVTR (g mm/m^2 day)	2.36
Oxygen transmission rate, OTR (cc mm/m^2 day)	55.12

Figure 5.19 Bicycle helmet made from knitted cellulosic high-performance fibers that had been used to reinforce Biopol B (9% 3-HV) [64].

diapers, feminine hygiene products, cosmetic containers, and cups as well as medical surgical garments, upholstery, carpet, compostable bags and lids, or tubs for thermoformed articles [67].

5.2.4.2 Polyhydroxybutyrate

PHB is the second member of the bacterial polyesters. It was synthesized from bacteria *Bacillus magaterium* by Lemoigne in 1925. PHBs are the class that is of interest for bio-derived and biodegradable plastics. The generalized chemical structure of PHB is:

$$-\{\ O-CH(CH_3)-CH_2-(C=O)\ \}_n\ -$$

PHB is completely biodegradable with a melting temperature (T_m) of 180°C, a glass transition temperature (T_g) of 5°C, and a high molecular

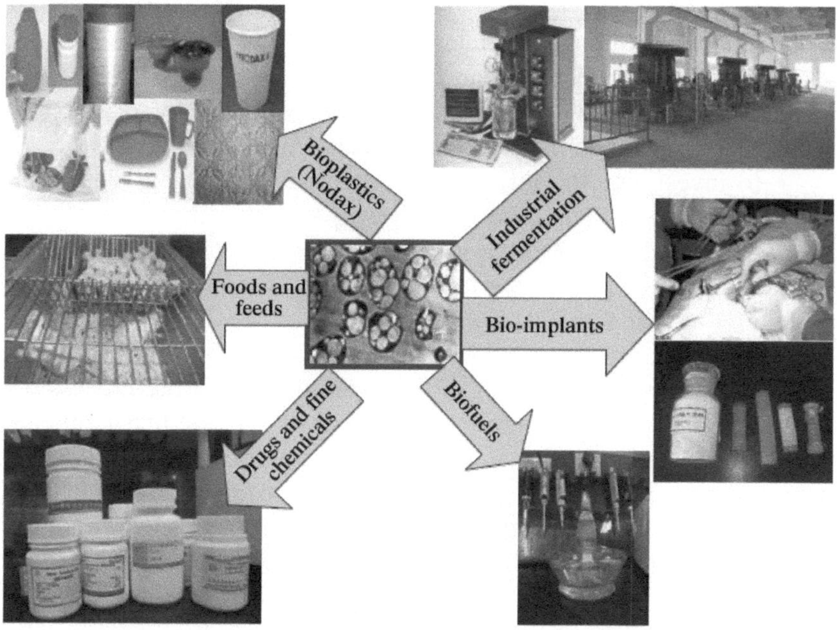

Figure 5.20 Applications of polyhydroxyalkanoates (PHAs) [67].

weight [68]. PHB is not crystalline in nature but it can be converted to more crystalline form by extraction process. Except for its biodegradability, PHB is similar to PP in the properties. PHB is more rigid and brittle and has very good barrier properties. The degree of crystallinity, glass transition, and microstructure of PHB dictates degree of brittleness in PHB. PHB tends to get brittle when it is stored for longer periods at room temperature. PHB can be easily processed as they are perfectly isotactic and does not include any chain branching. Unlike other synthetic polymers, they do not contain any residues of catalysts. Although PHBs are not water soluble, they are 100% biodegradable. Unlike other biodegradable polymers, PHBs are resistant to hydrolysis [69]. PHB has low permeability for oxygen, water, and carbon dioxide and resists oxidation. However, they only offer slight protection against chemical resistance.

For the development of a PHB biodegradable material with a better mechanical properties and long-term stability, it is necessary [68,69]:

- to prevent the degradation of chains in processing by addition of lubricants and plasticizers;
- to stop the secondary crystallization that occurs in the amorphous phase after aging time at room temperature;

- to investigate the effect of high temperatures melt processing, the cooling rate, and holding time on the crystallization rate, crystal size, and number of spherulites;
- to investigate the influence of the concentration of plasticizers on glass temperatures and other relaxation mechanisms;
- to investigate the influence of additives on biodegradation behavior of PHB (soil burial, river water, compost conditions, and the aerobic test);
- to optimize processing technology like, for instance, injection molding and extrusion.

Fig. 5.21 shows a flow chart of the PHB manufacturing process [68]. It is all initiated when sunlight converts carbon dioxide that is liberated from sugar beets or sugar cane into atmosphere to carbohydrates by photosynthesis. These carbohydrates are the raw material for the manufacture of PHB. PHB can be produced from glucose as a raw material or agricultural waste like, for instance, molasses or material, which is refined from the processing of sugar beets and lactose. The sugar splits up in the metabolism to C_2 building blocks, which are converted, over several steps, to C_4 monomers. Finally, PHB is polymerized.

The fermentation process as shown in Fig. 5.21 is carried out in two steps [68]:

1. The first step involves prefermentation. In this stage, the cells are grown at 30°C in a stream of air in a mineral medium with glucose as a carbon-source. The cells multiply after 24 h and reach approximately a cell density of 20 g/L.
2. During the second step, while fermentation process is on, the cells continue to multiply and after 40–80 h reach a cell density of 100 g/L. At this point, PHB occurs as discrete granules with diameters of (0.3–1 μm) in cells as storage material. At the end of the fermentation process, the cell has a dry mass 80% in the form of PHB.

Some of the commercialized PHB products available in the market are shown in Table 5.10 [24].

PHB is a versatile material. It is used as a surgical implant, in surgery, as seam threads for the healing of wounds and blood vessels. Because of its compatibility with the blood and tissues of mammals. PHBs are used as microcapsules in therapy or as materials for cell and

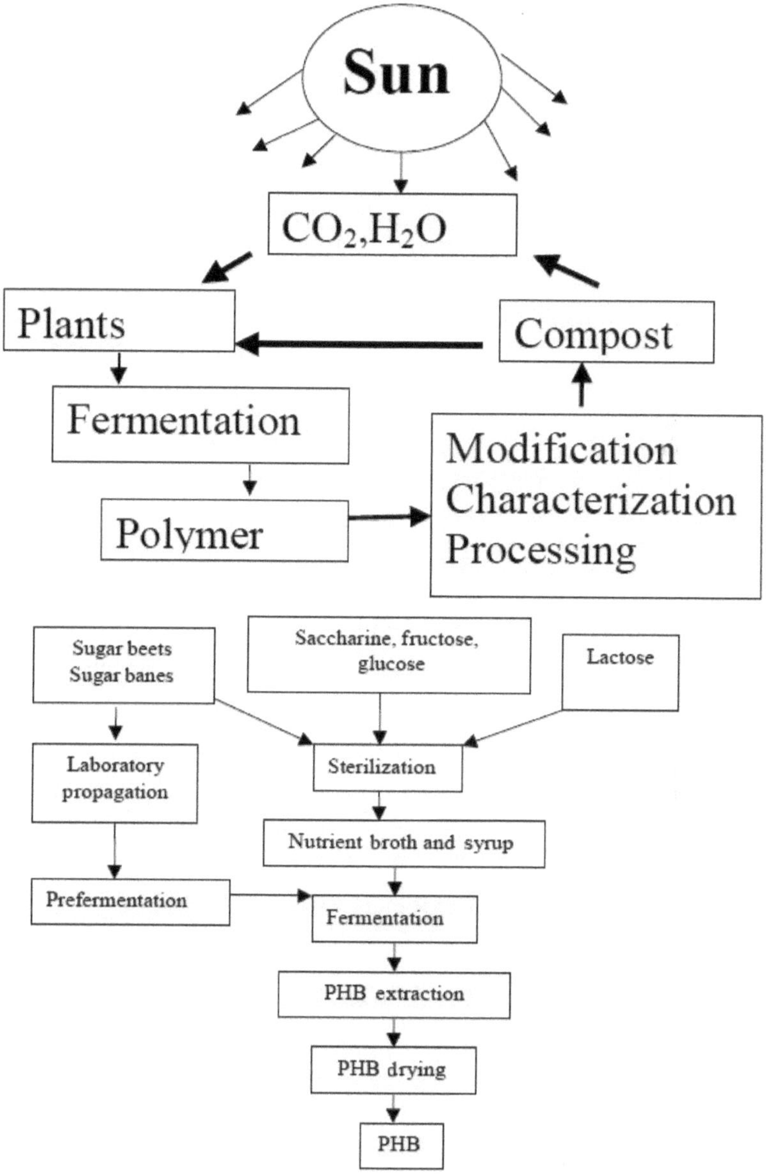

Figure 5.21 The life cycle of PHB [68].

tablet packaging. PHB is often used in packaging of deep draw articles for the food industry, for example, bottles, laminated foils, fishnets, potted flower, hygiene, fast food, one-way cups, agricultural foils, and fibers in textile [24].

Table 5.10 Commercial PHB Products [24]

Trade name	Product highlights
Biomer	• Produced by Alcaligenes Latus • Pellets are commercialized for classical plastic transformation processes • Allows injection molding of thin-walled parts • Composting duration is about 2 months
Nodax	• Developed by Procter & Gamble (P&G) • The polymer degrades both anaerobically and aerobically, has alkaline digestibility, and good surface properties for printing • Can be produced as fibers and nonwovens, foams, synthetic papers, films, and injection and blow molded articles
Biopol	• Developed by ICI, bought by Zeneca, then by Monsanto, and now the property of Metabolix. • It is a copolymer of hydroxybutyrate and hydroxyvalerate • Adapted for injection and blow molding, fibers, films, foaming, laminating, and thermoforming. • Good candidate for electric and electronic packaging applications because of its antistatic properties • Biopol is sensitive to hydrolysis despite its high degree of crystallinity

5.3 Synthetic Biodegradable Polymers

Aliphatic polyesters are among the most used synthetic biodegradable polymers in medical applications. All of the aliphatic polyesters have hydrolyzable ester linkages in their structure. Table 5.11 shows list of synthetic biodegradable polymers. In this category, polyglycolide, and polylactide, are often referred to as poly(glycolic acid) (PGA) and poly(lactic acid) (PLA). However, high-molecular-weight polymers of glycolic and lactic acid are not possible to obtain by direct condensation of the related carboxylic acids because of the reversibility of the condensation reaction, backbiting reactions, and the high extent of reaction required. Therefore, polyglycolide and polylactide are typically made by ring opening polymerization of their respective cyclic diester dimers, glycolide, and lactide [70].

Table 5.11 Synthetic Biodegradable Polymers [70]

Polymer	Structure
Poly(ε-caprolactone) (PCL)	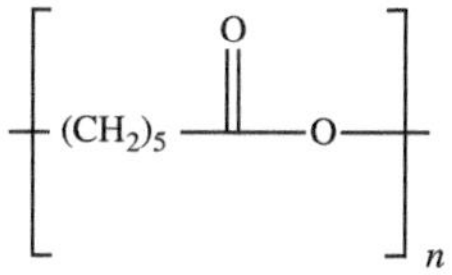
Polylactide or Poly(lactic acid) (PLA)	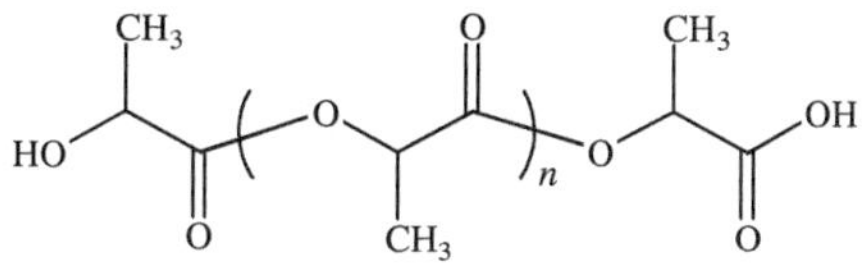
Polyglycolide or Poly (glycolic acid) (PGA)	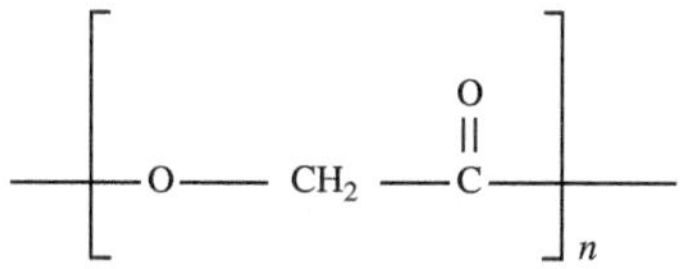
Poly(*p*-dioxanone) (PDS)	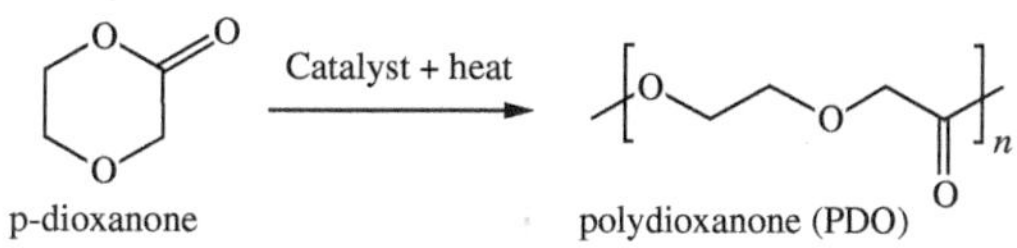
Poly(lactide-*co*-glycolide)	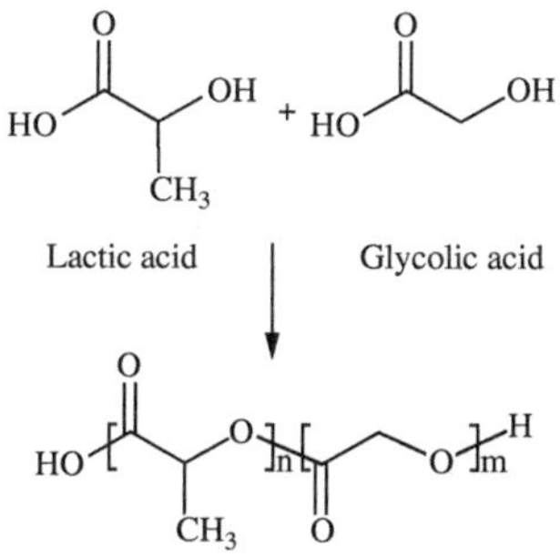

In theory, these aliphatic biodegradable polymers can be synthesized by the step growth condensation polymerization of either hydroxyacids [71]:

$$HO-R-COOH \longrightarrow -\!\!\left[O-R-CO\right]_{n}\!\!- + n\,H_2O$$

or diols and diacids:

$$HO-R-OH + HOOC-R'-COOH \longrightarrow$$
$$\left[HO-R-O-CO-R'-CO \right]_n + nH_2O$$

However, due to difficulty in removing the small molecule by product like water, the direct polycondensation will yield low-molecular-weight polyesters. High-molecular-weight aliphatic polyesters can be synthesized by ring opening polymerization of heterocyclic monomers bearing at least one ester bond in the monomer structure as shown in the reaction below [71].

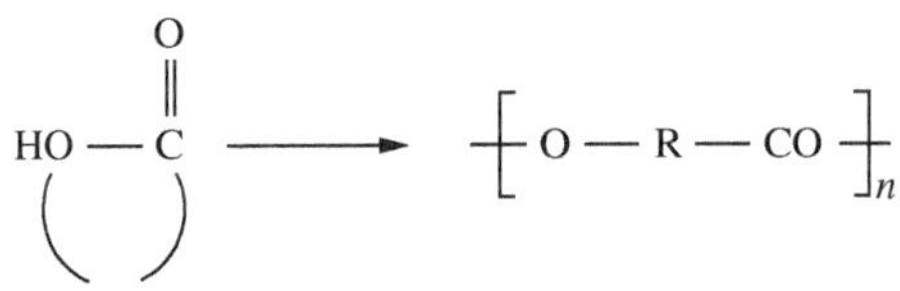

5.3.1 Poly(lactic acid)

PLA belongs to the family of aliphatic polyesters that are derived from renewable sources, mainly starch and sugar. It is a rigid thermoplastic polymer that can be semicrystalline or amorphous, depending on the stereochemistry of the polymer backbone. Lactic acid (2-hydroxy propionic acid), the building block of PLA can exist either in D- or L-enantiomers. The properties of PLA will depend on the proportion of the enantiomers allowing production of PLA with wide range of properties to match with the performance requirements [72].

PLA is considered very versatile because it behaves like polyethylene terephthalate (PET) and performs like PP. Its ability to be stress and thermally crystallized, impact modified, and copolymerized, PLA can have broadest range of applications. PLA can also be processed in most polymer processing equipment. Fig. 5.22 shows applications of PLA. It can be formed into transparent films, fibers, or injection blow molded preforms for bottles, like PET. PLA also has excellent organoleptic characteristics and is excellent for food contact and related packaging applications [72].

In comparison with petroleum-based polymers, PLA has reasonably good optical, physical, mechanical, and barrier properties [73]. The permeability coefficients of carbon dioxide, oxygen, nitrogen, and water for PLA are lower than that for PS, but higher than PET [73–75]. PLA has similar barrier properties to PET against organic permeants, such as ethyl acetate and D-limonene [76]. Although random-oriented PLA are brittle

Figure 5.22 Applications of PLA [72].

but they tend to possess good strength and stiffness. The performance of oriented PLA exceeds performance of oriented PS but it is comparable to oriented PET [9]. Tensile and flexural properties of PLA are higher than high-density polyethylene (HDPE), PP, and PS, but the Izod impact strength and elongation at break values are smaller than those for these polymers [77].

Lactic acid, the fundamental building block of PLA is produced by carbohydrate fermentation process or by chemical synthesis. However, most of lactic acid currently being produced is through fermentation. Fig. 5.23 shows different techniques such as azeotropic dehydrative condensation, direct condensation polymerization, and/or polymerization through lactide formation to produce high-molecular-weight PLA [78,79]. However, currently most of the high-molecular-weight PLA is produced by lactide ring opening polymerization [79–81].

Depending on the composition of the D- and L-enantiomers, PLA can crystallize in three forms. Thermal, barrier, and mechanical properties depends on the optical purity of PLA [82–88]. PLA with L-content greater than ~90% tend to be crystalline while those with lower optical purity are amorphous. Moreover, melting temperature (T_m), glass transition temperature (T_g), and crystallinity decreases with decreasing L-content [86,89]. PLA is known to undergo thermal degradation when exposed to elevated temperatures leading to the formation of lactide monomers [90–94].

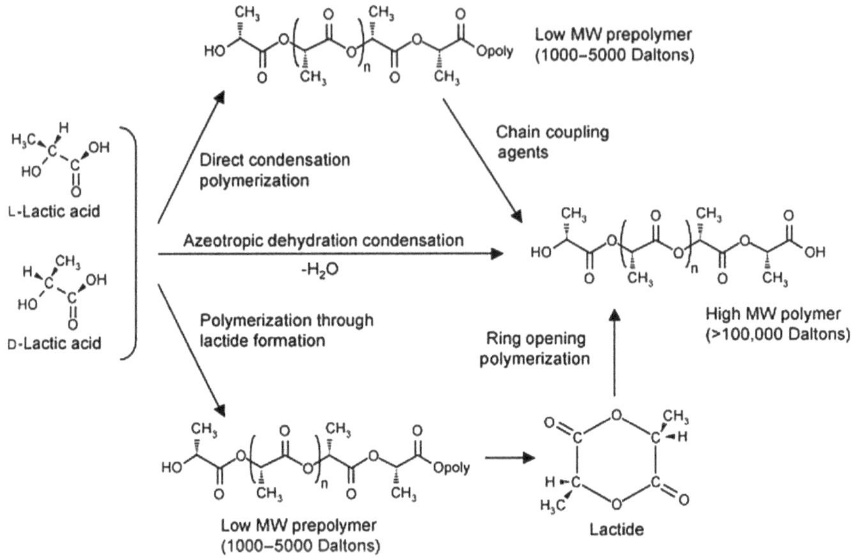

Figure 5.23 Synthesis of PLA from L- and D-lactic acids [78,79].

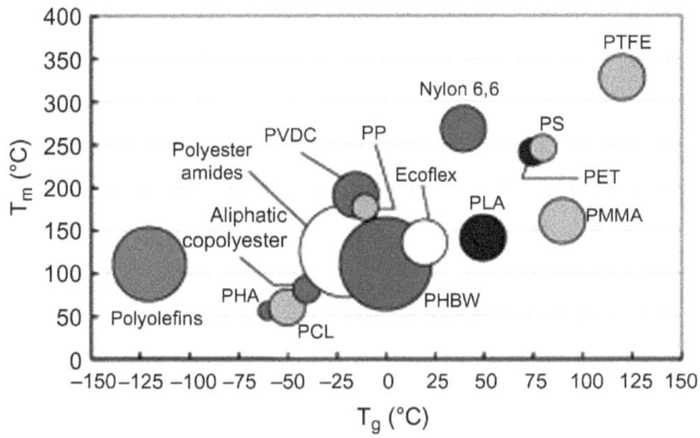

Figure 5.24 Comparison of glass transition and melting temperatures of PLA with other thermoplastics [87].

Similar to many thermoplastic polymers, semicrystalline PLA exhibits T_g and T_m. Above T_g ($\sim 58°C$) PLA is rubbery, while below T_g, it becomes a glass which is still capable to creep until it is cooled to its transition temperature at approximately $-45°C$, below which it behaves as a brittle polymer [95].

Fig. 5.24 shows comparison of glass transition temperature (T_g) and melting temperatures (T_m) of PLA with other polymers. In comparison to other polymers, PLA has relatively high T_g and low T_m [87].

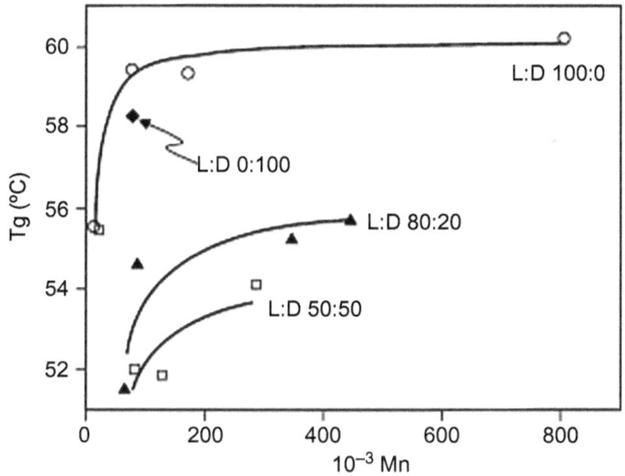

Figure 5.25 Glass transition temperatures for PLAs of different L-contents as a function of molecular weight [89].

Fig. 5.25 compares glass transition temperature of PLA with different optical purity of the polymer with respect to their molecular weights. It clearly shows that the T_g of PLA is dependent on both the molecular weight and the optical purity of the polymer. The glass transition temperature increases with molecular weight while PLA with higher content of L-lactide has higher T_g values than the same polymer with the same amount of D-lactide [89].

PLA tends to undergo thermal degradation in the molten state. Most of this form of degradation is related to processing of PLA which include process temperatures and the residence time in the extruder [96]. Other factors that can contribute to thermal degradation of PLA include [97]:

- hydrolysis by trace amounts of water;
- depolymerization;
- oxidative, random main-chain scission;
- intermolecular transesterification to monomer and oligomeric esters;
- intramolecular transesterification resulting in formation of monomer and oligomer lactides of low molecular weight.

Fig. 5.26 shows thermal degradation mechanism for PLA. Depending on the point in the backbone at which the reaction occurs, the product can be a lactide molecule, an oligomeric ring, or acetaldehyde plus carbon monoxide [98,99].

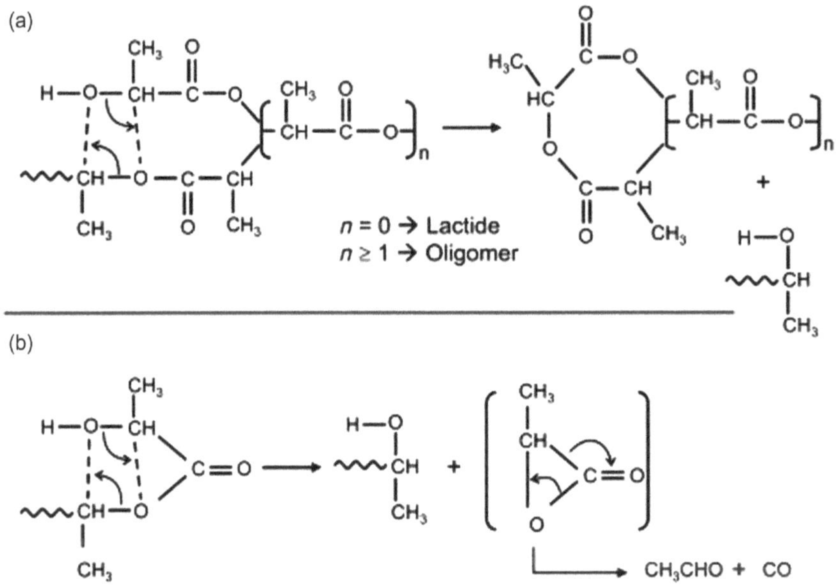

Figure 5.26 Thermal degradation of PLA [98,99].

PLA can be processed by extrusion, thermoforming, injection, blow molding, fiber spinning, and stretching. Most of its applications are found in food industry and geared toward packaging applications like packaging for food and drinks. It is also used for hygienic products for nonwoven materials. Its biocompatibility and bioresorption property makes them a good candidate for suture threads and clips, orthopedic fixations (screws, pins), and of resorbable implants. PLA is printable and sealable. Table 5.12 shows list of industry leaders for PLA market.

5.3.2 Poly(ε-caprolactone)

Poly(ε-caprolactone) PCL is a semicrystalline, biodegradable polymer with melting temperature (T_m) of ~60°C and a glass transition temperature (T_g) of ~ − 60°C. The structure and properties of PCL are shown in Fig. 5.27 [100].

PCL is synthesized by the ring opening polymerization of the cyclic monomer ε-caprolactone as shown in Fig. 5.28 [101]. Molecular weight of the polymer is controlled by low-molecular-weight alcohols while catalysts such as stannous octoate are used to catalyze the polymerization reaction. Polymerization of PCL can be affected by various mechanisms such as anionic, cationic, coordination, and radical mechanism.

Table 5.12 Main Products of Polylactic Acid (PLA)

Lacear (Mitsui Toatsu Chemical, Japan)	• Pellets exist for injection, molding of foam products, blow molding, thermoforming, extrusion, and fiber spinning
Eco-Pla – NatureWorks (Cargill Dow Polymers, USA)	• Dedicated to packaging applications films, thermoformed and injected products, paper coatings • The degradation duration is about 4–6 weeks • Products include fibers for clothes, hygiene products, or carpets
Lacty (Shimadzu Corporation, Japan)	• Produces PLA pellets for film and fiber extrusion
Lactron (Kanebo Goshen, Japan)	• Dedicated to the production of nets used in agriculture or for fishing • Exists also as a nonwoven product used in hygiene products
Solanyl (Rodenburg Biopolymers, Netherlands)	• PLA is made here from potatoes • Grades for injection are commercially available injection molding thin-wall parts
Galactic (Belgium)	• Commercialized for the production of films and fibers

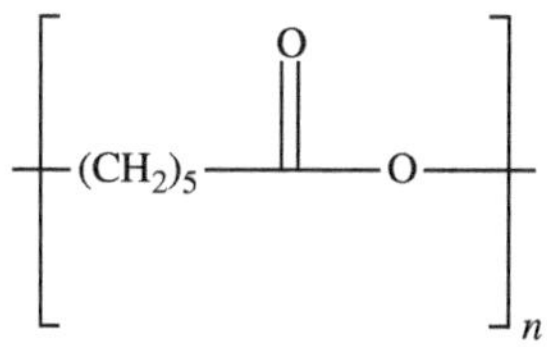

Figure 5.27 Structure of Poly(ε-caprolactone) (PCL) [100].

Each mechanism will have an effect on molecular weight, molecular weight distribution, end-group composition, and chemical structure of the copolymers [102].

Poly(ε-caprolactone) (PCL) is soluble in different solvents such as chloroform, dichloromethane, carbon tetrachloride, benzene, toluene, cyclohexanone, and 2-nitropropane at room temperature. However, it is

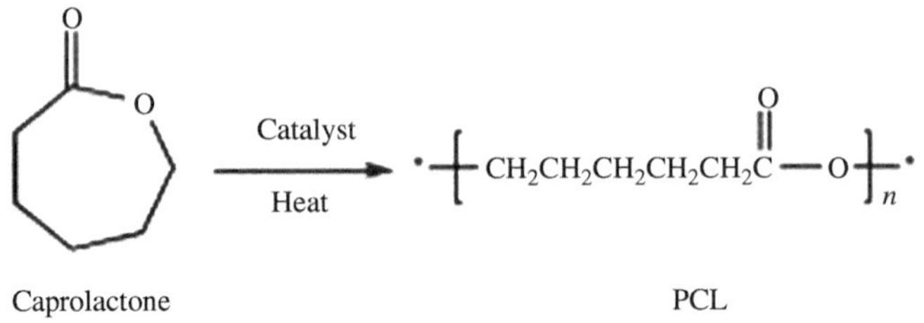

Figure 5.28 Synthesis of PCL [101,103].

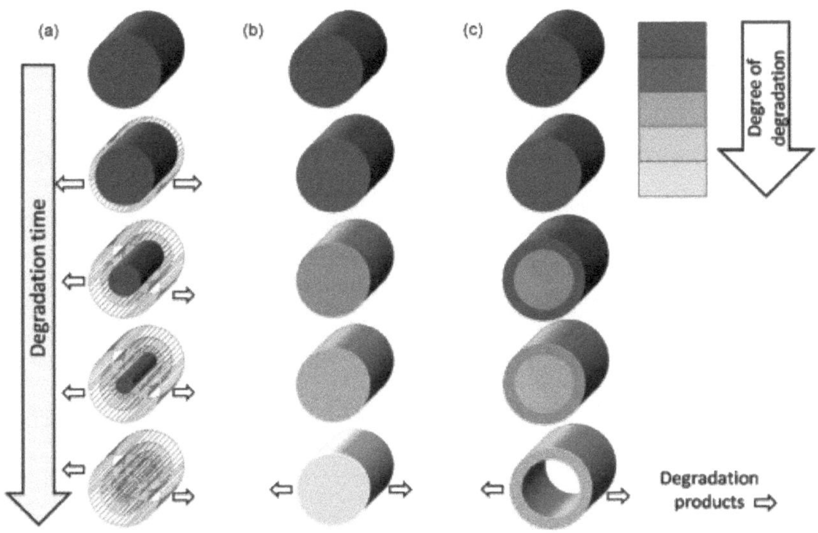

Figure 5.29 Degradation timeline when Poly(ε-caprolactone) (PCL) undergoes degradation using (a) surface erosion, (b) bulk degradation, (c) bulk degradation with autocatalysis [106].

least soluble in acetone, 2-butanone, ethyl acetate, dimethylformamide, and acetonitrile and is insoluble in alcohol, petroleum ether, and diethyl ether [104]. In order to improve stress crack resistance, dyeability, and adhesion, PCL are blended with other polymers. Some of these polymers include cellulose propionate, CAB, PLA, and Polylactic acid-co-glycolic acid [54].

Poly(ε-caprolactone) (PCL) undergoes biodegradation by outdoor living organisms such as bacteria and fungi; however, they do not degrade in animal and human bodies because they lack suitable enzymes [105]. Fig. 5.29a, b, and c shows degree of degradation of Poly(ε-caprolactone) (PCL) when it is exposed to three different mechanisms [106]. In the first mechanism, PCL undergoes degradation by surface erosion as shown in

Fig. 5.29a. The surface erosion involves the hydrolytic cleavage of the polymer backbone only at the surface [106,107]. This happens only when the rate of hydrolytic chain scission and the production of oligomers and monomers are faster than the rate of water intrusion into the polymer bulk, leading to thinning of polymer overtime without affecting the molecular weight [106].

Fig. 5.29b studies degradation timeline of PCL when it undergoes bulk degradation [106]. Bulk degradation happens when water penetrates into the entire polymer bulk, resulting in hydrolysis throughout the entire polymer matrix. This starts random hydrolytic chain scission which causes an overall reduction in molecular weight. If water diffuses into the polymer bulk and hydrolyzes the chains enabling the monomers or oligomers to diffuse out; thus leading to gradual erosion of polymer chain [106].

Fig. 5.29c shows degradation timeline of PCL when it undergoes bulk degradation using auto catalysis [106]. In this mechanism, the internal concentration of autocatalysis products produces an acidic gradient as the newly generated carboxyl end group formed during ester bond cleavage accumulates. This in turn accelerates the internal degradation compared to the surface, leaving an outer layer of higher molecular weight skin with a lower molecular weight, degraded, interior [106].

Due to high permeability, excellent biocompatibility toward many drugs, and its ability to be fully excreted from the body once bioresorbed, Poly(ε-caprolactone) (PCL) is suitable for controlled drug delivery applications. In comparison to other polymers, biodegradation of Poly(ε-caprolactone) (PCL) is slow which makes it most suitable for long-term delivery [108–110]. PCL are also used in sutures, wound dressings, contraceptive devices, fixation devices, and dentistry as shown in Fig. 5.30 [111,112].

5.3.3 Poly(glycolide) or Poly(glycolic acid)

Poly(glycolide) also known as PGA is a highly crystalline, biodegradable polymer having a melting point of ~225°C and a glass transition temperature of ~35°C. The molecular and crystalline structure of PGA allow very tight chain packing and thus have some very unique chemical, physical, and mechanical properties to the material. Because of its high degree of crystallization, it is not soluble in most organic solvents except for highly fluorinated organic solvents such as hexafluoroisopropanol [113]. The heat of fusion of 100% crystalline PGA is 45.7 cal/g. The structure of PGA is shown in Fig. 5.31 [113].

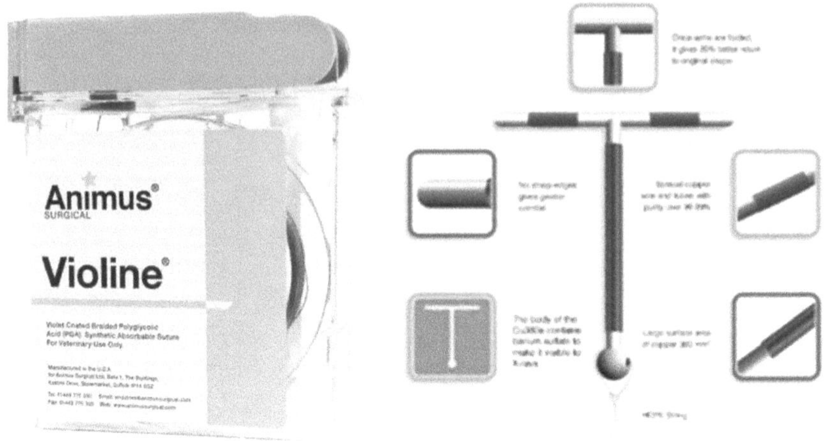

(a) (b)

Figure 5.30 Applications of PCL—(a) The suture is coated in PCL and calcium stearate and (b) Intrauterine device (IUD) [111,112].

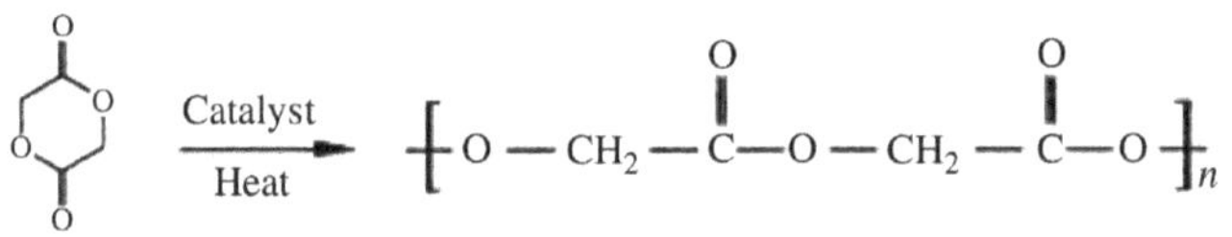

Figure 5.31 Structure of poly(glycolic acid) [113].

PGA is produced through the ring opening polymerization of glycolide which yields high-molecular-weight materials. The glycolide monomer is synthesized by dimerization of glycolic acid. The repeating molecular structure of PGA consists of one nonpolar methylene group and a single relatively polar ester group.

PGA fibers have exceptionally high strength and modulus because of its high degree of crystallinity as shown in Kuredux PGA data sheet in Fig. 5.32 [114]. However, PGA readily undergoes hydrolysis of the readily accessible and hydrolytically unstable aliphatic-ester linkages. The degradation time is just a few weeks depending on the molecular weight, degree of crystallinity, crystal morphology, and physical geometry.

PGA is a highly stiff material which makes it unsuitable to be used as sutures with an exception as braided material. Glycolide can be copolymerized with other monomers, such as lactide and trimethylene carbonate, to yield copolymer with various mechanical and biodegradation properties for different applications. PGA is also used in implantable medical devices

Properties		UNIT	VALUE	TEST METHOD
Melt index		g/10min	22~6	ISO 1133 Load:2.16kgf Temperature:250°C
Melt Viscosity		Pa•s	360~950	270° C, 122sec⁻¹
Density		g/cm3	1.5-1.6	Density Gradient Column Method
Thermal properties	Melt Temperature	° C	225	DSC:Initial Temperature:0° C Final Temperature:250° C Heating rate:10° C/min Purge Gas:N2 50ml/min
	Crystallization Temperature	° C	95	
	Glass Transition Temperature	° C	40	
	Thermal Expansion	1/K	5.40E-05	ASTM D696
	Thermal Conductivity	W/m•K	0.35	ASTM C177
	Combustion Heat	KJ/g	12	JIS-M 8814-1996
	Specific Heat	J/g•° C	1.12	JIS-K 7123
Optical properties	Refractive index		1.45-1.51	ASTM D542
	Haze	%	<1.0	JIS-K 6714
Physical properties*	Tensile Strength	Mpa	220	ASTM D882, 23° C-50%RH
	Elongation	%	25	ASTM D882, 23° C-50%RH
	Young's Modulus	GPa	6.5	ASTM D882, 23° C-50%RH
Barrier properties*	20° C-0%RH	cc/m²•day•atm	0.9	ISO 14663-2
O2 Transmission Rate	20° C-80%RH		0.9	
	20° C-90%RH		1.1	
Water Vapor Transmission Rate	40° C-90%RH	g/m²•day	11	JIS-Z 0208

*Biaxial-oriented film(400%x400%), 15um thick

The figures above are the result of lab measurements and are not secured values

Figure 5.32 Kuredux PGA data sheet [114].

including anastomosis rings, pins, rods, plates, and screws [115,116]. PGA have also been explored for tissue engineering, primarily in making scaffolds.

The Kureha Chemical Industries has commercialized high-molecular-weight PGA for food packaging applications under the trade name of Kuredux [117,118]. PGA is an exceptional barrier material because of its high degree of crystallization behavior. Therefore, the high molecular version of PGA is frequently used as an intermediate layer between PET layers in barrier films protection for perishable foods, including carbonated beverages and foods that lose freshness on prolonged exposure to air. Thinner plastic bottles which still retain desirable barrier properties may also be enabled by this polyglycolide interlayer technology. DuPont Co. has developed a low molecular weight version PGA which has an application in oil and gas industry [119] (Fig. 5.33).

5.3.4 Poly(*p*-dioxanone)

PDS is prepared by the ring opening polymerization of *p*-dioxanone shown in Fig. 5.34 [121]. It is a relatively weak, rapidly biodegrading polymer that was introduced into the market in the 1980s as a new

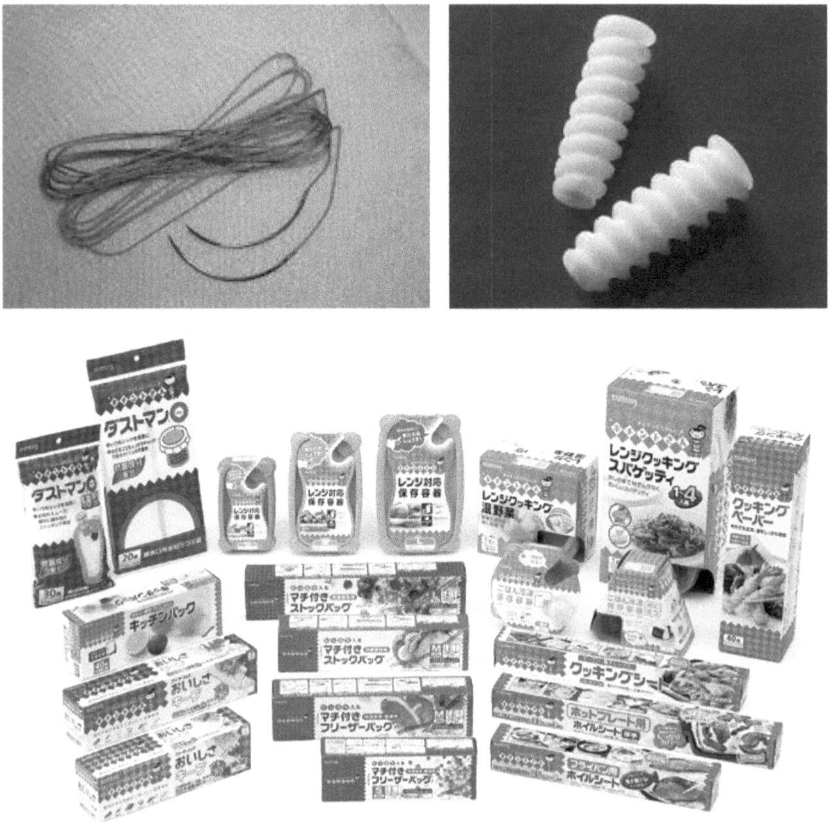

Figure 5.33 Some applications of poly(glycolic acid) [115,116,120].

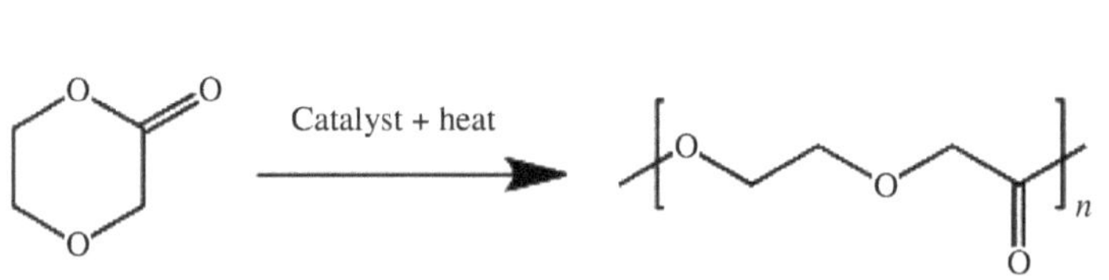

p-Dioxanone Polydioxanone (PDO)

Figure 5.34 Structure of poly(*p*-dioxanone) [121].

degradable suture material. PDS releases degradation products that are less acidic than the degradation products released by PGA or PLA [121].

PDS has about 55% crystallinity with a melting point of 106–115°C and glass transition temperature of −10°C to 0°C. In comparison to PGA, PDS is more flexible because of the ester linkage in its structure. The degradation of PDS is significantly slower both in vivo and in vitro

Table 5.13 Commercial Products of Poly(*p*-dioxanone) [113]

Product	Device	Processing method	Polymer type
PDS II	Suture	Extrusion	PDS homopolymer
BIOSYN	Suture	Extrusion	Terpolymer
ABSOLOK	Ligating clip	Injection molded	PDS homopolymer
LAPRA-TY	Suture clip	Injection molded	PDS homopolymer
ORTHOSORB	Pin	Extrusion	PDS homopolymer

because it contains fewer ester linkages in its molecular structure than PGA and PGLA copolymers. The degradation rate of PDS can be enhanced by incorporating of 5–25% glycolide as comonomer.

Although, PDS is relatively weak and lacks the stiffness and strength required for most orthopedics application, there has been a lot of research to find a solution. Development of small bone pins for the fixation of fractures in nonload bearing bones was the outcome of extensive research. Bone pins based on PDS are now marketed under the trade name Orthosorb in the United States and under the name Ethipin in Europe. However, its range of applications is limited. Academic institutions are also investigating PDS mostly as a polymer for implantable drug delivery applications [121,122].

PDS can also be injection molded into a number of nonfilamentous surgical devices, namely ABSOLOK and LAPRA-TY. Table 5.13 shows commercial products which utilize PDS [113].

5.3.5 Poly(lactide-co-glycolide)

Poly(lactide-*co*-glycolide) (PLGA) is a linear copolymer that can be prepared at different ratios between its constituent monomers, lactic acid (LA) and glycolic acid (GA) as shown in Fig. 5.35 [123,124].

Depending on the ratio of lactide to glycolide used for the polymerization, different forms of PLGA can be obtained. The ratio 75:25 identifies a copolymer consisted of 75% lactic acid and 25% glycolic acid [125]. PLGA can be synthesized using different mechanisms and the physical and chemical properties of PLGA are strongly influenced by the process parameters.

Low-molecular-weight PLGA is synthesized by polycondensation of lactic acid (LA) and glycolic acid (GA) at temperatures above 120°C under water removal conditions [126]. High molecular PLGA is

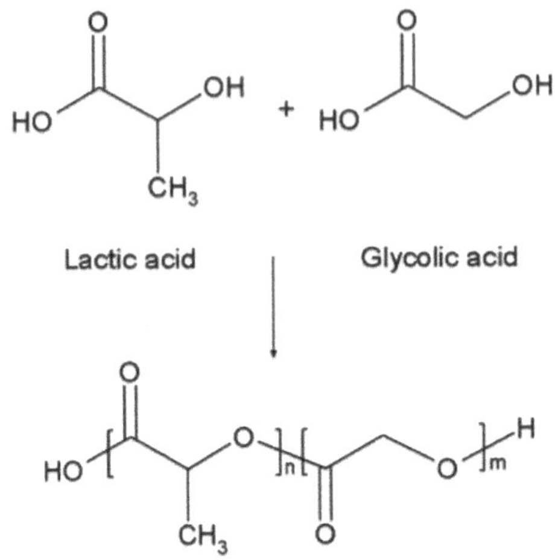

Figure 5.35 Chemical structure of poly(lactide-*co*-glycolide acid) and its monomer [123,124].

produced by a ring opening polymerization of lactide and glycolide, the cyclic diesters of lactic acid and glycolic acid respectively under catalyst such as stannous octanate. The reaction rate and final copolymer weight is controlled by adding a compound containing free hydroxy group which acts as a chain regulator [127].

Polymerization conditions such as comonomer ratio, catalyst concentration, temperature, and time can be varied to get products with varying yield, molar mass, and copolymer composition. The reaction rate of glycolide is much faster than that of lactide and this can lead to copolymers with glycolide block sequences along the polymer backbone [128].

PLGA can be dissolved by a wide range of common solvents, including chlorinated solvents, tetrahydrofuran, acetone, or ethyl acetate unlike polylactic and PGA which show poor solubility. PLGA can be processed into any shape and size, and can encapsulate biomolecules of any size [129]. Physical properties of PLGA depend on the initial molecular weight of the monomers, the ratio of lactic acid to glycolic acid, the exposure time to water and the storage temperature [130].

Some of the typical applications of PLGA in biomedical industry are:

- Porous scaffolds
- Fibrous scaffolds
- Hydrogels
- Injectable microspheres.

PLGA is also widely used in bone tissue engineering due to its suitable physical and chemical properties and biodegradability rate; however, compared to natural polymers such as collagen and chitosan which have numerous ionic molecular groups, the synthetic polymers have relatively few of these groups, making it very difficult to induce mineralization [131,132].

5.4 Bio-Derived Polyethylene

PE is an important polymer traditionally produced from fossil resources. PE is produced by polymerization of ethylene under pressure, temperature, in the presence of a catalyst. Traditionally, ethylene is produced through steam cracking of naphtha or heavy oils or ethanol dehydration. PE, which is one of the most common plastics, is a polymer of long chains of the monomer ethylene. There are several different types of PE but most common types include:

- LDPE—Low-density polyethylene
- HDPE—High-density polyethylene

PE is not biodegradable, leading to environmental issues associated with its use. Recycling of PE is relatively straight forward, although it would be considered to be environmentally advantageous to use a biodegradable alternative. The basic monomer of the polymer PE is ethylene. Ethanol which is very similar chemical product to ethylene can be produced by the fermentation of agricultural feedstocks such as corn or sugar cane—hence the expression Bio-PE.

Ethanol produced by fermentation from renewable resources can be used as a bio-fuel but also as a raw material for Bio-PE production. Production of Bio-PE from renewable resources consists of several steps:

1. Synthesis of ethanol by fermentation process from sugars, extracted from natural materials for example, sugarcane;
2. Chemical dehydration reaction transforming of ethanol into ethylene;
3. Classical reaction of polymerization of ethylene into PE.

Bio-PE is chemically and physically identical to traditional PE. Although it is not biodegradable it can be recycled. It can also considerably reduce greenhouse gas emissions because the technology of using it is made from sugar cane ethanol. For example, 1 ton production of Bio-PE captures, and removes from the environment, 2.5 tons of carbon dioxide, while the traditional petrochemical PE results in emissions close to 3.5 tons [133].

Bio-Low density polyethylene (Bio-LDPE) is produced by a process of addition polymerization. Bio-LDPE uses the process of addition polymerization to convert ethanol to Bio-PE product. The process starts when ethanol is produced by fermenting a biomass material such as corn starch or cane sugar is converted to ethylene which is then polymerized by addition polymerization into Bio-PE [134,135].

In 2010, Brazilian petrochemicals and polymers group Braskem started supplying their first Bio-PE. Braskem claimed that the production of every ton of their bio-based PE resulted in emission of 2.5 tons of carbon dioxide which was much less when compared to the traditional petrochemical method where emissions were close to 3.5 tons [136]. A typical of its process is shown in Fig. 5.36 [137].

Number of leading companies have started developing Bio-PE products and establishing themselves as major players in this area. Some of the companies and their achievements are discussed in this section [135]:

1. AT&T

   AT&T is the first US telecom company to use packaging that is composed of up to 30% plant-based materials sourced from sugarcane ethanol. AT&T has started using Bio-PE in packaging wireless accessories, which includes most device cases and power accessories as shown in Fig. 5.37 [138].

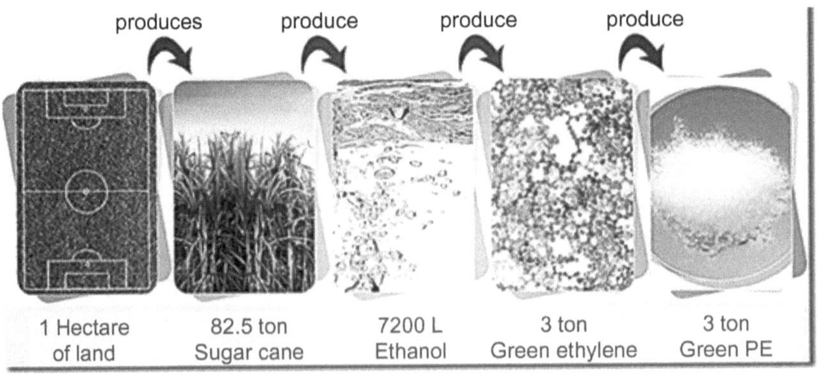

Figure 5.36 Braskem's typical Bio-PE production process [137].

Figure 5.37 Bio-derived polyethylene packaging used by AT&T for their wireless accessories [138].

2. Braskem

   Brazilian national petrochemical giant, Braskem have invested approximately $290 million to produce 200,000 metric tons of sugarcane-based PE annually in Brazil. Braskem currently produces line or products "I'm green" which are PE produced from ethanol sugarcane. Fig. 5.38 shows some of Braskem's Bio-PE commercial applications [139,140].

3. Ecover

   Ecover, the green cleaning brand uses Bio-PE obtained from sugarcane in all of its packaging known as Plantastic [141]. Fig. 5.39 shows some of its examples [142].

4. Johnson & Johnson

   Johnson & Johnson in Brazil launched Brazilian sunscreen line, Sundown, in a new packaging that contains sugarcane-based PE from Braskem. The package contains 60% green plastic and 40% recycled material in Fig. 5.40 [143].

5. Nestlé & Tetra Pak

   Nestlé Brazil, in partnership with Tetra Pak and Braskem have started using Bio-PE produced from sugarcane to make caps for box-type packages of two popular Nestlé milk brands, Ninho, and Molico shown in Fig. 5.41 [144,145].

Figure 5.38 Some of the commercial applications of Braskem's bio-derived polyethylene [139,140].

6. Pantene

   Pantene is using Bio-PE in its new shampoo and conditioner bottles. The plant-based plastic is featured on the Pantene Pro-V Nature Fusion collection shown in Fig. 5.42 [146].

Figure 5.39 Ecover's plantastic applications [142].

Figure 5.40 Johnson & Johnson's sunscreen line, Sundown in bio-derived polyethylene packaging [143].

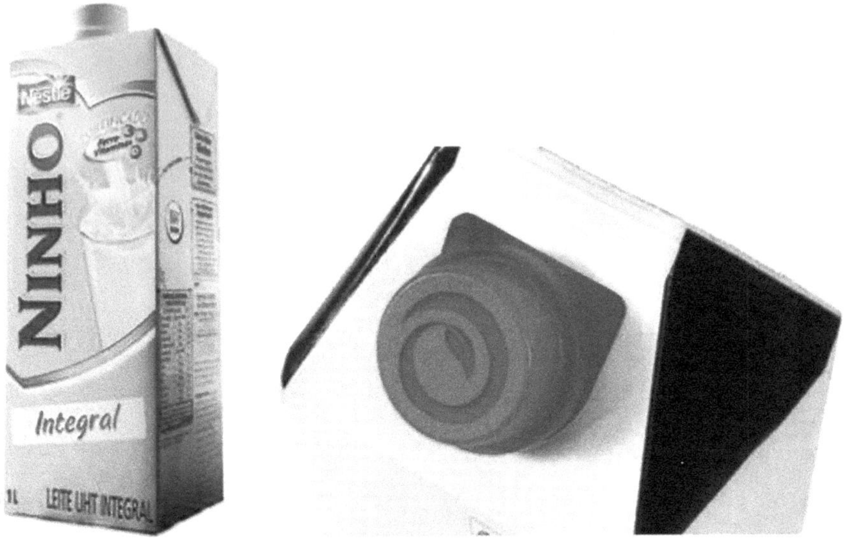

Figure 5.41 Cap's made from bio-derived polyethylene [144,145].

Figure 5.42 Pantene Pro-V nature fusion shampoo and conditioners [146].

5.5 Polymers from Bio-Based Monomers

Bio-based materials are produced primarily by two different ways: (1) the direct production of polymers and (2) the production of bio-based monomers and their further polymerization. However, the main difficulty is to trigger its composition. For example, in the case of PHA, different microorganisms can produce the same polymer; however, modification of the feeding substrates will impact the monomeric composition of the polymers produced by the same microorganism. The main difficulty consists in making the microorganism perform the biosynthesis consistently and repeatedly. This issue can be resolved by more controlled polymerization with bio-based monomers.

The production of bio-based monomers can help resolve the issue of making microorganism perform the biosynthesis consistently and repeatedly. However, the polymerization of bio-based monomers often requires more development, as in the case of PLA and of PBS. There are two more options that can be foreseen [147]:

1. The production of partially bio-based materials (Sorona)
2. The production of bio-monomers identical to the already existing and improved petroleum-based (ethylene, isobutylene, caprolactam, etc.)

In this category, following bio-based monomers will be discussed:

1. Lactic acid and Poly(Lactic Acid) (PLA)
2. 1,3 propanediol and DuPont's Sorona
3. Succinic acid
4. Bio-based unsaturated polyester (UPE).

5.5.1 Lactic Acid and Poly(lactic acid)

Lactic acid has been an important molecule used in food and chemical industries. Typically, it is produced through anaerobic fermentation by many different bacteria. It has been used as a natural acidifying agent in food industry [148,149]. More recently, lactic acid has been used in the synthesis of PLA as a new biodegradable and bio-based polymer [150].

The synthesis of PLA from lactic acid cannot really be achieved by simple condensation, which mainly leads instead to the oligomers. Therefore, PLA can be achieved by ring opening polymerization reaction as shown in Fig. 5.43. This ring opening polymerization reaction starts from the lactide, the lactic acid dimer, with the lactide itself being obtained by the partial depolymerization of the oligomers [151–154].

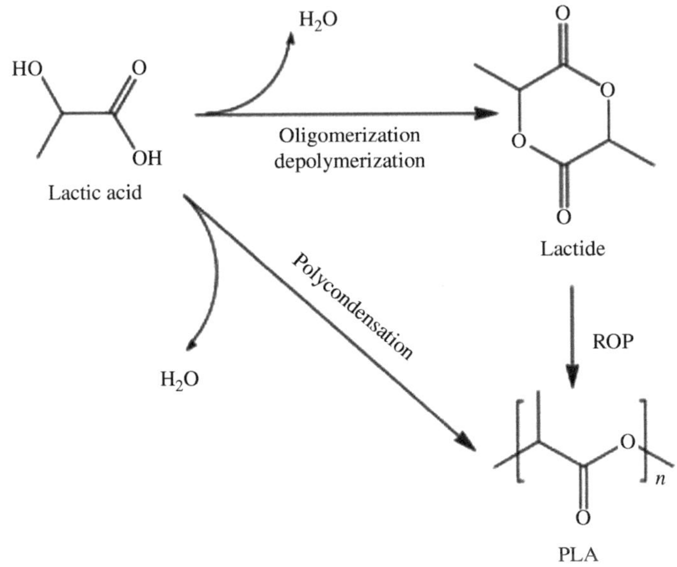

Figure 5.43 Synthesis of PLA [151].

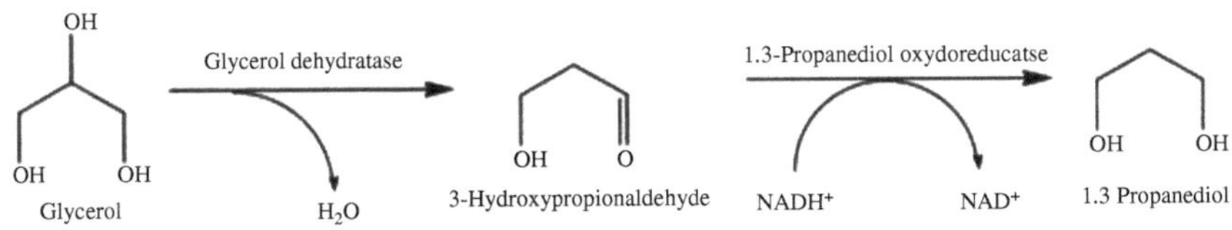

Figure 5.44 Metabolic pathway for the production of 1,3-propanediol from glycerol [147].

5.5.2 1,3 Propanediol and DuPont's Sorona

1,3-Propanediol (PDO) is one of the oldest known products of anaerobic fermentation. Fig. 5.44 shows metabolic pathway for the production of PDO from glycerol [147]. The production of PDO by wild-type strains leads to the production of several by-products, such as 2, 3-butanediol, lactate, acetate, formate, and ethanol.

DuPont performed the polymerization of PDO with terephthalic acid to obtain Sorona and is shown in Fig. 5.45 [155]. A further simplified process is shown in Fig. 5.46a [156].

This partially bio-based polymer is mainly used in different fiber applications such as floor coverings and sportswear shown in Fig. 5.46b [157].

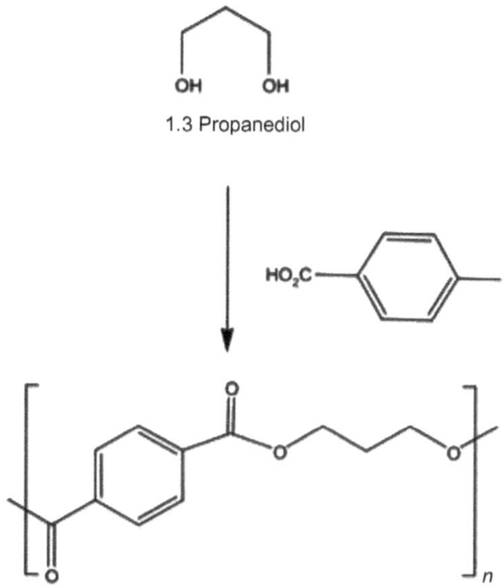

Figure 5.45 Synthesis of Sorona [155].

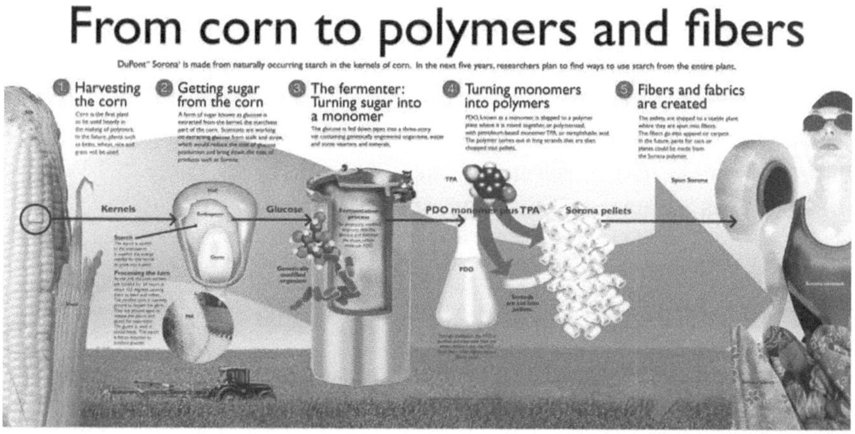

(a)

(b)

Figure 5.46 (a) Simplified process to manufacture DuPont Sorona, and (b) its use in flooring applications [156,157].

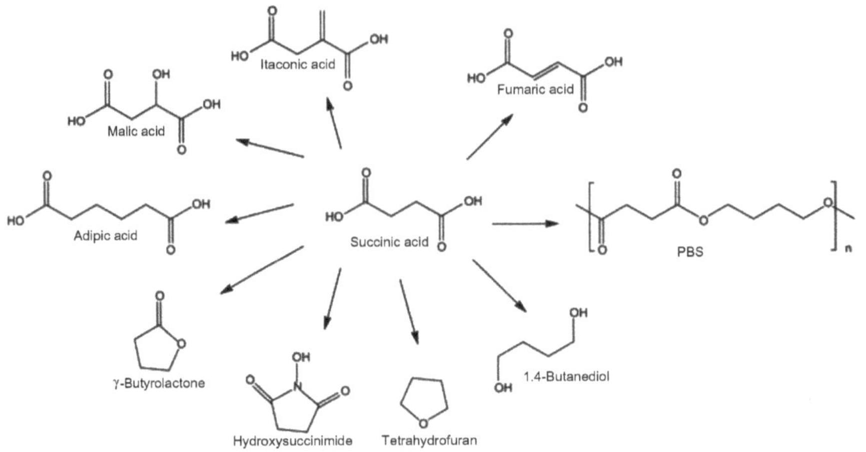

Figure 5.47 Different products of succinic acid [161].

5.5.3 Succinic Acid

Succinic acid is also produced by anaerobic fermentation of several wild-type strains [158–160]. Succinic acid is considered as one of the major bio-based platform chemicals and lead to different products as shown in Fig. 5.47 [161].

Succinic acid is used in a variety of applications in (1) food, (2) pharmaceuticals, (3) agricultural, and (4) industrial. Table 5.14 shows different uses for each application [162].

5.5.4 Bio-Based Unsaturated Polyester

UPEs are a large family of polymers with a wide range of applications and their ability to undergo various postpolymerization reactions. They can be utilized in high-gloss coatings, insulating materials, drug delivery systems, and biomedical applications such as tissue engineering [163–164].

The synthesis of a range of UPEs from bio-derived molecules can be seen in Fig. 5.48. All diesters and diols and polyols used are obtainable from biomass via thermal, chemical, or biological treatment of biomass within a biorefinery facility [165].

Plant-oil based UPEs are typically produced by the condensation of carboxylic acid with an alcohol. Thus, esters are functional groups derived from carboxylic acid, where the hydroxyl (OH) group is replaced with an alkoxy group (OR) group. Plant oil is an important

Table 5.14 Different Applications of Succinic Acid [162]

Area of application	Uses
Food	• As a flavoring agent in food and beverages • For dry gelatin desserts, for preparing cake flavorings • As bread softening agents • As sodium succinate as flavor enhancer • As catalyst for food seasoning preparation • In the synthesis of modified starch • As micro encapsulation of flavoring oils • For protein gelatination • In the preservation of chicken against *Salmonella* • In the synthesis of modified starch • Dog food preservative
Pharmaceuticals	• In the preparation of active calcium succinate, which acts as anticarcinogenic agent and as deodorant for removal of fish odor • Starting material for Active Pharmaceutical Ingredients (APIs) including adipic acid, *N*-methyl pyrrolidinone, 2-pyrrolidinone, succinate salts, 1,4-butanediol, tetrahydrofuran, and γ-butyrolactone • As an additive in formulation, as an insulinotropic agent, and in drug control release polymers • Medicines of sedative, antispasmer, antiplegm, antiphogistic, anrhoter, contraception, and cancer curing • In the preparation of Vitamin A • In the preparation of Antiinflammatory—Erthyrodiol derivates • Antidote for toxic substance
Agriculture	• Seed treatment to enhance germination, growth, and yield • Plant growth regulator and accelerator • Root stimulator for grapes • Soil chelating agent • Crop dust antistat • Cut flower preservative • Component of barley seed antismut treatment

Table 5.14 Different Applications of Succinic Acid [162]—cont'd

Area of application	Uses
Industrial	• For curing epoxy resins, polyester synthesis, and other applications • As an intermediate in the manufacture of dyes, perfumes, lacquers, photographic chemicals, alkyd resins, plasticizer, metal treatment chemical, vehicle water cooling systems, alkyd resins, and coatings and succinates • Plasticizer for polymers • Biodegradable solvents and lubricants • Engineering plastics • Adhesive and powder coating • Corrosion inhibitor

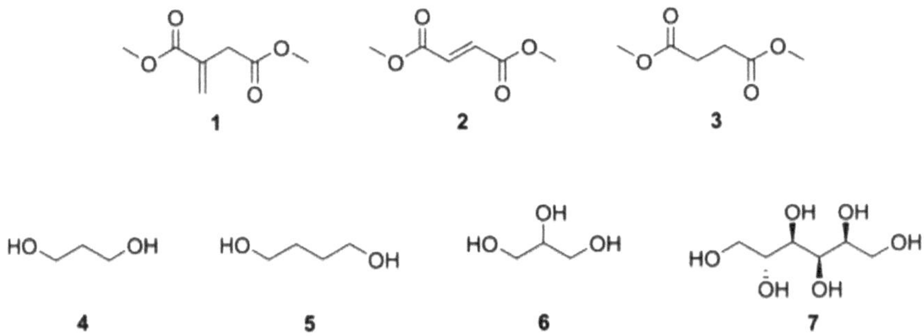

Figure 5.48 Bio-derived platform molecules (1–3) and various polyols (4–7) used for synthesis of unsaturated polyesters (UPEs). (1) Dimethyl itaconate (DMI), (2) dimethyl fumarate (DMFu), (3) dimethyl succinate (DMS), (4) 1,3-propanediol, (5) 1,4-butanediol, (6) glycerol, and (7) sorbitol. Conditions: Ti(IV)O^{t}Bu$_4$ catalyst, 175–200°C, 29.5–68 h [165,166].

source that can be used as a precursor to obtain polyester because the functional groups present in triglyceride molecules, such as double bond, hydroxyl, epoxy, ester, acid, etc., can provide a great variety of bio-based polymers, with different structures and functionalities, via a desired chemical reaction [167].

Fig. 5.49 shows polyesters formed by condensation reaction between a diacid and a dialcohols, with elimination of one water molecule. The polycondensation reaction occurs through step growth polymerization.

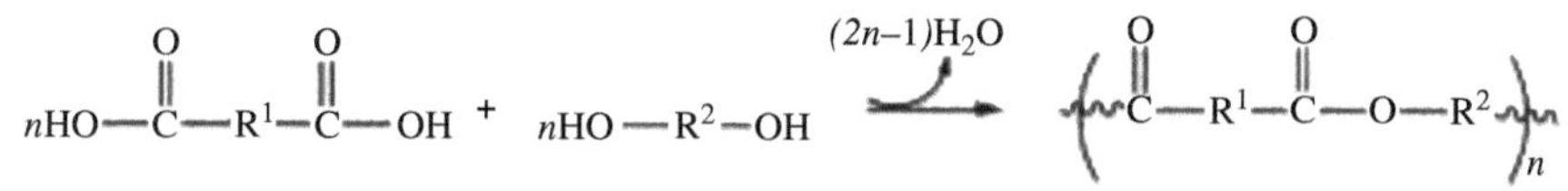

Figure 5.49 Polycondensation reaction to obtain polyester [167].

The *mers* unit R1 and R2 are the repeating units in the polyester main chain, and they can drive the applications depending on the used monomers [167].

DuPont's Susterra PDO is used as the glycol portion in the manufacturing of UPE resins where it has been formulated and commercialized globally in a wide variety of products. For manufacturers looking to increase the bio-based content of their resins, Susterra offers a 100% renewably sourced product that can easily be used to replace part or the entire glycol component used in the polyester reaction. These resins incorporating Susterra propanediol have found use in the construction, marine, and automotive industry worldwide with end users that require high performance polymers with a more sustainable footprint [168].

5.6 Bio-Based Materials for Paper/Substrate Coatings

In recent years, there has been a growing interest in the use of bio-based resin building blocks in the synthesis of polymers for use in coatings. Bio-based products are derived from plants and other renewable agricultural, marine, and forestry materials and provide an alternative to conventional petroleum-derived products [169].

Driving forces include a growing public and private awareness and interest in the use of renewable raw materials that can meet sustainability expectations and certifications such as Green Seal and Green Guard as well as the USDA BioPreferred Program for product labeling. Green Seal and Green Guard are environmentally driven, whereas the USDA BioPreferred program functions to encourage the use of renewable agricultural raw materials in products [169]. ASTM D6866 was developed to standardize, certify, and classify the bio-based content of materials. Minimum renewable carbon content categories (MRCC) have been established for purchasing by Federal agencies and their contractors as shown in Table 5.15 [170].

Bio-based resins for coatings are normally referred to as alkyds. Alkyds are comprised of fatty acid modified polyester resins. These resins are

Table 5.15 BioPreferred Coating Categories [170]

Coating and paint category	MRCC (%)
Interior Latex and Waterborne	20
Oil-based and solvent borne alkyd	67
Wood and concrete stains	39
Others	25

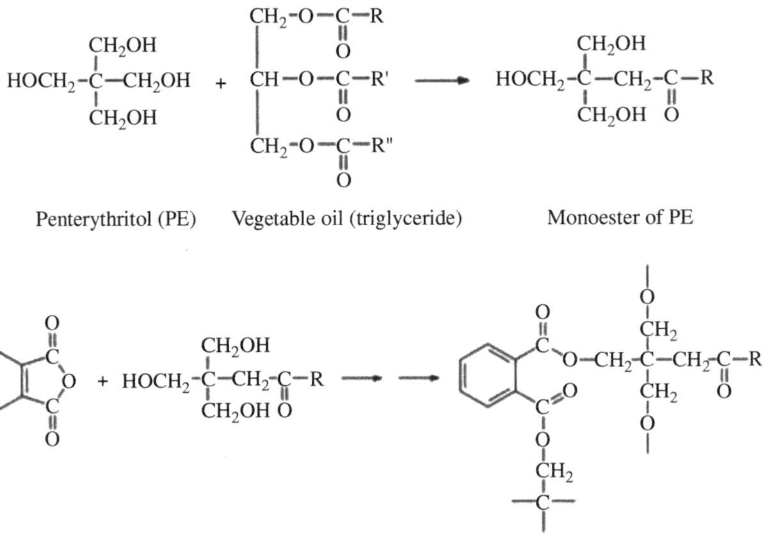

Figure 5.50 Schematics of a simplified reaction sequence to form an alkyd resin from penterthritol, a vegetable oil, and phthalic anhydride [171].

sometimes modified to include a urethane linkage and thus called uralkyds or oil modified urethanes. The fatty acid portion is derived from naturally occurring or renewable oils derived from sunflower, safflower, soybean, castor, tall, and others. Polyesters are derived from the reaction product of a polyol and a di or multifunctional acid or carboxylic acid and anhydride to form multiple ester linkages in a polymer chain.

Newer bio-based resin technologies also feature low odor, low or no VOC, free of alkoxy phenolethoxylates (APE) with performance comparable to their solvent borne counterparts. The challenge is to formulate bio-based resins and paints that provide performance attributes equal or better than that derived from coatings formulated with higher levels of petroleum-based materials. Fig. 5.50 shows schematics of a simplified reaction sequence to form an alkyd resin from penterthritol, a vegetable oil, and phthalic anhydride [171].

Table 5.16 Examples of Bio-Based Building Blocks for Resin Synthesis [172–174]

Natural oil	Source	Acid
Castor	Castor oil seeds	Biosuccinic acid
Coconut	Coconut kernels	Biooctadecanoic acid
Linseed	Linseed seeds	Bio 2,5-furandicarboxylic acid
Safflower	Safflower seeds	Bioazelaic acid
Sunflower	Sunflower seeds	Bio 3-hydroxypropionic acid
Tung	Tung tree	Bioacrylic acid
Tall	Wood pulp	

Table 5.16 shows examples of bio-based building blocks for resin synthesis [172–174].

Bio-based resins containing a high level of MRCC polymer building blocks currently in the market include PUDs based on polyols derived from castor oil. Renewable waterborne alkyd latexes are reported to have the flow leveling characteristics of their solvent borne counterpart. Soy oil and castor oil based polyols are also available. Caprolactone lactide copolymers can be optimized for use as polyols in coatings.

The use of bio-based raw materials in resins and coatings is a growing area of interest as many companies are adding such materials to their product portfolio. Properly formulated bio-based coatings can provide equivalent performance to their nonbio-based counterparts providing low or no VOC, low odor, APE free, and with MRCC content to be qualified as BioPreferred coatings. Manufacturers of bio-based raw materials and resins include Alberdingk Boley, Cargill, Purac, Perstorp, and BYK USA (EU).

References

[1] B. Ghanbarzadeh, H. Almasi, Biodegradable polymers, in: R. Chamy, F. Rosenkranz (Eds.), Biodegradation – Life of Science, InTech Publishers, 2013, p. 141 (Chapter 6).

[2] B. Anne, Biodegradable polymers and composites, Integrated Waste Management, vol. 1, InTech Publishers, 2011, p. 341 (Chapter 18).

[3] A. Gregorová, Classification and Physical Properties of Biodegradable Polymers, Physical Properties of Biodegradable Polymers, Institute for Chemistry and Technlogy of Materials, Graz University of Technology (Chapter 2).

[4] L. Averous, E. Pollet, Environmental Silicate Nanobiocomposites, Green Energy and Technology, Springer, Hiedelberg, 2012.
[5] R. Babu, K. O'Connor, R. Seeram, Current progress on bio-based polymers and their future trends, Prog. Biomater. 2 (8) (2013), Springer.
[6] Bio-based Polymers, Chemical & Engineering News, 92 (43) (2014) pp. 10–14. <http://cen.acs.org/articles/92/i43/Biobased-Polymers.html#>.
[7] R. Pareta, M.J. Edirisinghe, A novel method for the preparation of starch films and coatings, Carbohydr. Polym. 63 (2006) 425–431.
[8] D.R. Lu, C.M. Xiao, S.J. Xu, Starch-based completely biodegradable polymer materials, Expr. Polym. Lett. vol. 3 (No. 6) (2009) 366–375.
[9] P. Tomasik, C.H. Schilling, Chemical modification of starch, Adv. Carbohydr. Chem. Biochem. 59 (2004) 175–403.
[10] M. Ramesh, J.R. Mitchell, S.E. Harding, Amylose content of rice starch, Starch 51 (1999) 311–313.
[11] R.A. Wallace, J.L. King, G.P. Sanders, Biology—The Science of Life, Goodyear Publishing Company, California, 1981.
[12] D.K. Platt, The Starch-Based Biodegradable Polymer Market – Applications and Market, iSmithers Rapra Publishing, Shawbury, UK, 2006.
[13] Making materials. <http://www.epnoe.eu/polysaccharides/making_materials>.
[14] M. Visser, EcoGen: biodegradable starch based polymer. <http://www.design-4-sustainability.com/materials/39-ecogen-biodegradable-starch-based-polymer>.
[15] J. Gotro, Thermoplastic Starch: A renewable biodegradable bioplastic, Polymer Innovation Blog. <http://polymerinnovationblog.com/thermoplastic-starch-a-renewable-biodegradable-bioplastic>.
[16] Vegetarian wound dressing, Community Research and Development Information Service (CORDIS). <http://cordis.europa.eu/result/rcn/82273_en.html>.
[17] Water Lock Superabsorbent Polymers, Grain Processing Corporation. <http://www.pharmaceuticalonline.com/doc/water-lock-superabsorbent-polymers-0001>.
[18] Biodegradable Plastic: Its Promises and Consequences. <http://dujs.dartmouth.edu/applied_sciences/biodegradable-plastic-its-promises-and-consequences#.VdPyAE3bKUk>.
[19] A.A. Curvelo, A.J.F. de Carvalho, J.A.M. Agnelli, Thermoplastic starch-cellulosic fibers composites: preliminary results, Carbohydr. Polym. v. 45 (2001) 183–188.
[20] E. Rudnik, Compostable Polymer Materials, Elsevier, Oxford, UK, 2008.
[21] N. Lourdin, V. Della, P. Colonna, P. Poussin, Polymères biodègradables: mise en ceuvre et propriétés de l'amidon, Caoutchoucs et Plastiques, n. 780 (1999).

[22] O. Vilpoux, L. Averous, Starch-based plastics, technology, use and potentialities of Latin American Starchy Tubers, in: M.P. Cereda, O. Vilpoux (Eds.), Collection Latin American Starchy Tubers, NGO Raízes and Cargill Foundation, São Paolo–Brazil, Book No. 3, 2004, pp. 521–553 (Chapter 18).

[23] L. Averous, C. Fringant, L. Moro, Starch-based biodegradable materials suitable for thermoforming packaging, Starch 53 (8) (2001) 368–371.

[24] R. Smith, Biodegradable polymers for industrial applications, CRC Press, Boca Raton, 2005.

[25] NIIR Board of Consultants and Engineers, The Complete Book on Biodegradable Plastics and Polymers, Asia Pacific Business Press Inc, Delhi, India, 2006, p. 6.

[26] S. Pilla, Handbook of bioplastics and biocomposites engineering applications, Wiley Publishers, Hoboken, NJ, 2011.

[27] L.A. Utracki, Role of polymer blends technology in polymer recycling, Polymer Blends Handbook, Springer, Netherlands, 2002.

[28] X. Tang, S. Alavi, Recent advances in starch, polyvinyl alcohol based polymer blends, nanocomposites and their biodegradability, Carbohydr. Polym. 85 (2011) 7–16.

[29] J.-M. Lagarón (Ed.), Multifunctional and Nanoreinforced Polymers for Food Packaging, first ed., Elsevier, Cambridge, UK, 2011.

[30] Sketch the cellulose structure. <http://www.biochemcs.com/cellulose_structure.htm>.

[31] Cellobiose, European Molecular Biology Laboratory. <http://www.ebi.ac.uk/chebi/searchId.do?chebiId = CHEBI:17057>.

[32] The Merck Index, 8th ed., Merck & Co, Rahway, NJ, USA, 1968.

[33] Pyranose. <http://science.uvu.edu/ochem/index.php/alphabetical/o-p/pyranose/>.

[34] Cellulose. <http://www.fibersource.com/f-tutor/cellulose.htm>.

[35] P.M. Visakh, S. Thomas, Preparation of bionanomaterials and their polymer nanocomposites from waste and biomass, Waste Biomass Valorizat. 1 (1) (2010) 121–134.

[36] Y. Sawai, T. Nozoe, Y. Yoshitani, D. Tsukada, Development of new cellulose based polymers with excellent melt processability, Fujifilm Res. Dev. No. 57 (2012).

[37] Eastman Cellulose Esters, Technical Literature. <http://www.eastman.com/Literature_Center/P/PCI105.pdf>.

[38] D.L. Kaplan, Biopolymers From Renewable Resources, Springer, Berlin, Germany, 1998.

[39] O. Olabisi, K. Adewale, Handbook of Thermoplastics, CRC Press, Boca Raton, FL, 1997.

[40] E. Lokensgard, Industrial Plastics: Theory and Applications, fifth ed., Delmar Cengage Learning, Boston, MA, 2010.

[41] Cellulose Acetate. <http://wpage.unina.it/avitabil/testi/AcetilCellulosa.pdf>.

[42] Manufacturing Process of Acetate Rayon. <http://textilelibrary.weebly.com/acetate-rayon-mfg-process.html>.

[43] M.M.L. Loo, R. Hashim, C.P. Leh, Recycling of valueless paper dust to a low grade cellulose acetate: effect of pretreatments on acetylation, BioResources 7 (1) (2012) 1068–1083<https://www.ncsu.edu/bioresources/BioRes_07/BioRes_07_1_1068_Loo_HP_Recycling_Paper_Dust_Cellulose_Acetate_Pretreat_2309.pdf>

[44] Cellulose Acetate Propionate, Sigma-Aldrich. <http://www.sigmaaldrich.com/catalog/product/aldrich/330183?lang=en&region=US>.

[45] Cellulose Acetate Butyrate, Sigma-Aldrich. <http://www.sigmaaldrich.com/catalog/product/aldrich/419044?lang=en&region=US>.

[46] J.E. Boyd, Celluloid: The Eternal Substitute. <http://www.chemheritage.org/discover/media/magazine/articles/29-3-celluloid-the-eternal-substitute.aspx>.

[47] J.A. Reilly, Celluloid objects: their chemistry and preservation, J. Amr. Inst. Cons. 30 (2) (1991) 145–162<http://cool.conservation-us.org/jaic/articles/jaic30-02-003.html>.

[48] S. Smith, The History and Properties of Celluloid. <http://www.ottawavalleybuttonclub.ca/Articles/tabid/544/articleType/ArticleView/articleId/409/The-History-and-Properties-of-Celluloid.aspx>.

[49] Encyclopedia Americana, Grolier, Danbury, CT, 1983 p. 303.

[50] G.S. Brady, H.R. Clauser, Materials Handbook, eleven ed, McGraw-Hill, New York, 1977.

[51] Worden, Nitrocellulose Industry, vols. 1 & 2, Van Nostrand, New York, 1911.

[52] L.S. Nair, C.T. Laurencin, Biodegradable polymers as biomaterials, Prog. Polym. Sci. 32 (2007) 762–798.

[53] C. Shi, Y. Zhu, X. Ran, M. Wang, Y. Su, T. Cheng, Therapeutic potential of chitosan and its derivatives in regenerative medicine, J. Surg. Res. 133 (2006) 185–192.

[54] R. Chandra, H. Rastogi, Biodegradable polymers, Prog. Poly. Sci. 23 (1998) 1273–1335.

[55] S.W. Shalaby, J.A. DuBose, M. Shalaby, Chitosan-based systems, in: S.W.B. Shalaby, J.L. Karen (Eds.), Absorbable and Biodegradable Polymers, CRC Press, Boca Raton, FL, 1977.

[56] H. Onishi, H. Takahashi, M. Yoshiyasu, Y. Machida, Preparation and in vitro properties of N-Succinylchitosan or carboxymethylchitin-mitomycin C conjugate microparticles with specified size, Drug Deliv. Ind. Pharm. 27 (2001) 659–657

[57] J.M. Pachence, M.P. Bohrer, J. Kohn, Biodegradable polymers, in: B. Lanza, C. Langer, P. Vacanti (Eds.), Principles of Tissue Engineering, third Ed., Elsevier, San Diego, 2007, p. 323.

[58] A.K. Mohanty, M. Misra, L.T. Drzal, Natural Fibers Biopolymers and Biocomposites, CRC Press, Boca Raton, FL, 2005.
[59] U.J. Hanggi, Requirements on bacterial polyesters as future substitute for conventional plastics for consumer goods, FEMS Microbiol Rev, 16, 1995, pp. 213–220.
[60] A.M. Clarinval, J. Halleux, Classification of Biodegradable Polymers, CRIF, France.
[61] What are PHAs, Renewable and Biodegradable Polymer for Engineered Solution, PolyFerm Canada. <http://www.polyfermcanada.com/pha.html>.
[62] A. Girdhar, M. Bhatia, S. Nagpal, A. Kanampalliwar, A. Tiwari, Process parameters for influencing polyhydroxyalkanoate producing bacterial factories: an Overview, J. Pet. Environ. Biotechnol. 4 (5) (2013).
[63] H.J. Endres, A. Siebert-Raths, Basics of PHA, Bioplast. Mag. vol. 6 (March 2011) 42–45<http://f2.hs-hannover.de/fileadmin/media/doc/ifbb/Bioplastics_Magazine__03_11__Vol._6_S._43-45.pdf>.
[64] B. Koch, Developpement d'un nouveau composite biodegradable: exemples concernant son procede de fabrication, son comportement lors de sa degradation et ses utilisations, Ph.D. thesis, Department of Material Science, Eidgenossische Technische Hochschule, Zurich, 1993.
[65] E. Wintermantel, M. Teichrnann, B. Koch, J. Mayer, Russ weg - Filter weg, Herstellung eines biologisch abbaubaren Filters, Umweltprojekte 92, Austellung uber Umwelttechnik and Umweltforschung an der Eidgenossischen Technischen Hochschule, Zurich, 1992.
[66] H. Brandl, R. Bachofen, J. Mayer, E. Wintermantel, Degradation and applications of PHA, Can. J. Microbiol. 41 (1) (1995) 143–153.
[67] C. Guoqiang, The application and industrialization of bio-based degradable materials – pHA, China Plast. Rubber J. (2010).
[68] Manufacturing and Properties of PHB. <http://sundoc.bibliothek.uni-halle.de/diss-online/02/02H017/t2.pdf>.
[69] A.M. Clarinval, J. Halleux, Classification of biodegradable polymers, in: R. Smith (Ed.), Biodegradable Polymers for Industrial Applications, CRC Press, Boca Raton, 2005 (Chapter 1).
[70] D. Shi, Introduction to Biomaterials, World Scientific Press, London, UK, 2011.
[71] J.B. Park, J.D. Bronzino, Biomaterials: Principles and Applications, CRC Press, Boca Raton, FL, 2002, pp. 55–77.
[72] Poly(lactic acid) (PLA) is Gaining Traction in the Market, Polymer Innovation Blog. <http://polymerinnovationblog.com/poly-lactic-acid-pla-is-gaining-traction-in-the-market/>.
[73] R. Auras, B. Harte, S. Selke, R. Hernandez, Mechanical, physical, and barrier properties of poly(lactide) films, J. Plast. Film Sheeting 19 (2003) 123–135.

[74] R. Auras, B. Harte, S. Selke, Effect of water on the oxygen barrier properties of poly(ethylene terephthalate) and polylactide films, J. Appl. Polym. Sci. 92 (2004) 1790–1803.
[75] R.A. Auras, S.P. Singh, J.J. Singh, Evaluation of oriented poly(lactide) polymers vs. existing PET and oriented PS for fresh food service containers, Packag. Technol. Sci. 18 (2005) 207–216.
[76] H. Lehermeier, J. Dorgan, J.D. Way, Gas permeation properties of poly (lactic acid), J. Membr. Sci. 190 (2001) 243–251.
[77] R. Auras, R. Harte, S. Selke, Sorption of ethyl acetate and d-limonene in poly(lactide) polymers, J. Sci. Food. Agric. 86 (2006) 648–656.
[78] J.R. Dorgan, H. Lehermeier, M. Mang, Thermal and rheological properties of commercial-grade poly(lactic acid)s, J. Polym. Environ. 8 (2000) 1–9.
[79] L.T. Lim, R. Auras, M. Rubino, Processing technologies for polylactic acid, Prog. Polym. Sci. 33 (2008) 820–852.
[80] R. Auras, B. Harte, S. Selke, An overview of polylactides as packaging materials, Macromol. Biosci. 4 (2004) 835–864.
[81] D. Garlotta, A literature review of poly(lactic acid), J. Polym. Environ. 9 (2001) 63–84.
[82] Y. Ikada, H. Tsuji, Biodegradable polyesters for medical and ecological applications, Macromol. Rapid Commun. 21 (2000) 117–132.
[83] H. Urayama, S.I. Moon, Y. Kimura, Microstructure and thermal properties of polylactides with different l- and d-unit sequences: importance of the helical nature of the l-sequenced segments, Macromol. Mater. Eng. 288 (2003) 137–143.
[84] H. Tsuji, R. Okino, H. Daimon, K. Fujie, Water vapor permeability of poly(lactide)s: effects of molecular characteristics and crystallinity, J. Appl. Polym. Sci. 99 (2005) 2245–2252.
[85] J.R. Sarasua, A.L. Arraiza, P. Balerdi, I. Maiza, Crystallinity and mechanical properties of optically pure polylactides and their blends, Polym. Eng. Sci. 45 (2005) 745–753.
[86] D.M. Bigg, Effect of copolymer ratio on the crystallinity and properties of polylactic acid copolymers, Annu. Techn. Conf. Soc. Plast. Eng. (1996) 2028–2039.
[87] H. Tsuji, Y. Ikada, Crystallization from the melt of PLA with different optical purities and their blends, Macromol. Chem. Phys. 197 (1996) 3483–3499.
[88] M. Hutchinson, J. Dorgan, D. Knauss, S. Hait, Optical properties of polylactides, J. Polym. Environ. 14 (2006) 119–124.
[89] M. Pyda, R.C. Bopp, B. Wunderlich, Heat capacity of poly(lactic acid), J. Chem. Thermodyn. 36 (2004) 731–742.
[90] J.R. Dorgan, J. Jansen, M.P. Clayton, Melt rheology of variable l-content poly(lactic acid), J. Rheol. 49 (2005) 607–619.
[91] T. Tsukegi, T. Motoyama, Y. Shirai, Racemization behavior of l,l-lactide during heating, Polym. Degrad. Stab. 92 (2007) 552–559.

[92] Y. Fan, H. Nishida, Y. Shirai, T. Endo, Control of racemization for feedstock recycling of PLLA, Green. Chem. 5 (2003) 575–579.
[93] F.D. Kopinke, M. Remmler, K. Mackenzie, M. Moder, O. Wachsen, Thermal decomposition of biodegradable polyesters, II. Poly(lactic acid), Polym. Degrad. Stab. 53 (1996) 329–342.
[94] F. Khabbaz, S. Karlsson, A.-C. Albertsson, Py-GC/MS an effective technique to characterizing of degradation mechanism of poly(l-lactide) in the different environment, J. Appl. Polym. Sci. 78 (2000) 2369–2378.
[95] C. Westphal, C. Perrot, S. Karlsson, Py-GC/MS as a means to predict degree of degradation by giving microstructural changes modelled on LDPE and PLA, Polym. Degrad. Stab. 73 (2001) 281–287.
[96] D.E. Henton, P. Gruber, J. Lunt, J. Randall, Polylactic acid technology, in: A.K. Mohanty, M. Misra, L.T. Drzal (Eds.), Natural fibers, biopolymers, and biocomposites, Taylor & Francis, Boca Raton, FL, 2005, pp. 527–577.
[97] V. Taubner, R. Shishoo, Influence of processing parameters on the degradation of poly(l-lactide) during extrusion, J. Appl. Polym. Sci. 79 (2001) 2128–2135.
[98] A. Sodergard, M. Stold, Properties of lactic acid based polymers and their correlation with composition, Prog. Mater. Sci. 27 (2002) 1123–1163.
[99] I.C. McNeill, H.A. Leiper, Degradation studies of some polyesters and polycarbonates. 1. Polylactide: degradation under isothermal conditions, thermal degradation mechanism and photolysis of the polymer, Polym. Degrad. Stab. 11 (1985) 309–326.
[100] G.F. Moore, S.M. Saunders, Advances in Biodegradable Polymers, vol. 9, No. 2, iSmithers Rapra Publishing, Shawbury, UK, 1997.
[101] R.F. Storey, A.E. Taylor, Effect of stannous octoate concentration on the ethylene glycol-initiated polymerization of epsilon-caprolactone, Abstr. Pap. Am. Chem. Soc. 211 (1996) 114–120.
[102] M. Okada, Chemical syntheses of biodegradable polymers, Progr. Polym. Sci. 27 (2002) 87–133.
[103] S. Ebnesajjad, Handbook of Biopolymers and Biodegradable Plastics, Elsevier Publishing, Oxford, UK, 2013.
[104] O. Coulembier, P. Degee, J.L. Hedrick, P. Dubois, From controlled ring-opening polymerization to biodegradable aliphatic polyester: especially poly(beta-malic acid) derivatives, Progr. Polym. Sci. 31 (2006) 723–747.
[105] M. Vert, Degradable and bioresorbable polymers in surgery and in pharmacology: beliefs and facts, J. Mater. Sci. Mater. Med. 20 (2009) 437–446.
[106] M.A. Woodruff, D.W. Hutmacher, The return of a forgotten polymer – Poly(ε-caprolactone), Prog. Polym. Sci. 35 (2010) 1217–1256.

[107] R. Ginde, R. Gupta, In vitro chemical degradation of poly(glycolic acid) pellets and fibers, J. Appl. Polym. Sci. 33 (1987) 2411–2429.

[108] A. Merkli, C. Tabatabay, R. Gurny, J. Heller, Biodegradable polymers for the controlled release of ocular drugs, Progr. Polym. Sci. 23 (1998) 563–580.

[109] S. Freiberg, X. Zhu, Polymermicrospheres for controlled drug release, Int. J. Pharm. 282 (2004) 1–18.

[110] V.R. Sinha, K. Bansal, R. Kaushik, R. Kumria, A. Trehan, Poly-epsiloncaprolactone microspheres and nanospheres: an overview, Int. J. Pharm. 278 (2004) 1–23.

[111] Violine Cassettes, Animus Surgical. <http://www.animussurgical.com/product/violine-cassettes/ - the suture is coated in PCL and calcium stearate>.

[112] Intrauterine Devices (IUD), Meril Life Science Private Limited. <http://www.indiamart.com/meril-lifescience/endo-surgery.html>.

[113] D.E. Perrin, J.P. English, in: A.J. Domb, J. Kost, D. Wiseman (Eds.), Polyglycolide and polylactide, Handbook of Biodegradable Polymers, CRC Press, 1998, pp. 3–5 (Chapter 1).

[114] Kuredux: Polyglycolic Acid Resin, Kureha Corp. <http://www.kureha.com/pdfs/Kureha-KUREDUX.pdf>.

[115] Polyglycolide, Wikipedia. <https://en.wikipedia.org/wiki/Polyglycolide>.

[116] Implantable Polymer Compounds, Foster Delivery Science. <http://www.deliveryscience.com/sites/default/files/Implantable_Polymers_Brochure.pdf>.

[117] D.K. Gilding, A.M. Reed, Biodegradable polymers for use in surgery—polyglycolic/poly (lactic acid) homo- and copolymers, Polymer 20 (12) (1979) 1459–1464.

[118] J. Middleton, A. Tipton, Synthetic biodegradable polymers as medical devices, Med. Plast. Biomater. Mag. (1998).

[119] E. Schmitt, Polyglycolic acid in solutions, U.S. Pat 3737440 (1973).

[120] Kichinto-San Series, Kureha Corporation. <http://www.kureha.co.jp/en/business/polymer/kichinto.html>.

[121] Z. Zheng, O. Ortiz, R. Goyal, J. Kohn, Biodegradable Polymers, in: K. Modjarrad, S. Ebnesajjad (Eds.), Handbook of Polymer Applications in Medicine and Medical Devices, Elsevier Publishers, Oxford, UK, 2014, pp. 318–319 (Chapter 13).

[122] S.G. Kumbar, C.T. Laurencin, M. Deng, Natural and Synthetic Biomedical Polymers, Elsevier, Oxford, UK, 2014.

[123] M.N.V.R. Kumar, Nano and Microparticles as controlled drug delivery devices, J. Pharm. Pharm. Sci. 3 (2) (2000) 234–258.

[124] P. Gentile, V. Chiono, I. Carmagnola, P.V. Hatton, An overview of poly (lactic-co-glycolic) acid (PPLGA)-based biomaterials for bone tissue engineering, Int. J. Mol. Sci. 15 (3) (2014) 3640–3659.

[125] S.B. Zhou, X.M. Deng, X.H. Li, W.X. Jia, L. Liu, Synthesis and characterization of biodegradable low molecular weight aliphatic polyesters and their use in protein-delivery systems, J. Appl. Polym. Sci. 91 (2004) 1848–1856.

[126] Z.Y. Wang, Y.M. Zhao, F. Wang, J. Wang, Syntheses of poly(lactic acid-*co*-glycolic acid) serial biodegradable polymer materials via direct melt polycondensation and their characterization, J. Appl. Polym. Sci. 99 (2006) 244–252.

[127] H.R. Kricheldorf, C. Boettcher, K.U. Tonnes, Polylactones: Polymerization of *racemic* and *mes* dl-lactide with various organotin catalysts stereochemical aspects, Polymer 33 (1992) 2817–2824.

[128] A. Kowalski, A. Duda, S. Penczek, Mechanism of cyclic ester polymerization initiated with tin(II) octoate 2 macromolecules fitted with tin(II) alkoxide species observed directly in MALDI-TOF spectra, Macromolecules 33 (3) (2000) 689–695.

[129] H.K. Makadia, S.J. Siegel, Poly lactic-*co*-glycolic acid (PLGA) as biodegradable controlled drug delivery carrier, Polymers (Basel), 3, 2011, pp. 1377–1397.

[130] M.L. Houchin, E.M. Topp, Physical properties of PLGA films during polymer degradation, J. Appl. Polym. Sci. 114 (2009) 2848–2854.

[131] K. Kawasaki, M. Aihara, J. Honmo, S. Sakurai, Y. Fujimaki, K. Sakamoto, et al., Effects of recombinant human bone morphogenetic protein-2 on differentiation of cells isolated from human bone muscle and skin, Bone 23 (1998) 223–231.

[132] Z.X. Meng, H.F. Li, Z.Z. Sun, W. Zheng, Y.F. Zheng, Fabrication of mineralized electrospun PLGA and PLGA/gelatin nanofibers and their potential in bone tissue engineering, Mater. Sci. Eng. C 33 (2013) 699–706.

[133] Sugarcane products, Bioplastics. <http://sugarcane.org/sugarcane-products/bioplastics>.

[134] A.B. Koltuniewicz, Bioprocesses, Sustainable Process Engineering: Prospects and Opportunities, Walter de Gruyter GmbH, Germany, 2014, pp. 319–321 (Chapter 5).

[135] I. Voevodina, A. Krzan, Bio-based polymers, Plastice. <http://www.plastice.org/fileadmin/files/biobased.pdf>.

[136] Future prospects for industrial biotechnology, OECD, OECD Publishing (2011).

[137] J. Gotro, Bio Polyethylene: Drop-in Replacement, Polymer Innovation Blog. <http://polymerinnovationblog.com/bio-polyethylene-drop-in-replacement/>.

[138] P. Reynolds, AT&T is sweet on sugarcane-based thermoform, Greener package. <http://www.greenerpackage.com/renewable_resources/att_sweet_sugarcane-based_thermoform>.

[139] B. Bouckley, Tetra Pak unleashes plant power with world's first 100% bio-based carton, Beverage Daily. <http://www.beveragedaily.com/Processing-Packaging/Tetra-Pak-launches-world-s-first-100-plant-based-carton>.
[140] Plastico Verde. <http://tatirossi.com/Plastico-verde>.
[141] Ecover's 100% sustainable plastic packaging, Horizons. <http://horizons.innovateuk.org/case-studies/124>.
[142] Ecover. <http://your.ecover.com/articles/plant-plastic>.
[143] Kookykitsch.com. <https://kookykitsch.com/index.php/medicinal-advertising/johnson-johnson-sundown-sunscreen-34-oz-spf-6-vintage-80s-detail>.
[144] Nestle launches bioplastic caps for milk brands in Brazil, Bio-based News. <http://news.bio-based.eu/nestle-launches-bioplastic-caps-for-milk-brands-in-brazil/>.
[145] S. Orman, Tetra Pak introduces innovative sugar cane bottle cap. <http://www.fdfworld.com/production/46/Tetra-Pak-Introduces-Innovative-Sugar-Cane-Bottle-Cap>.
[146] A.M. Mohan, Sugarcane-based Pantene bottle to launch in Western Europe, greener Package. <http://www.greenerpackage.com/bioplastics/sugarcane-based_pantene_bottle_launch_western_europe>.
[147] N. Berezina, S.M. Martelli, Bio-based polymers and materials, Renewable Resources for Biorefineries, RSC Green Chemistry, Cambridge, UK, 2014, pp. 1−28 (Chapter 1). <http://pubs.rsc.org/en/content/chapterhtml/2014/bk9781849738989-00001?isbn=978-1-84973-898-9#sect160>.
[148] Y.J. Wee, J.N. Kim, H.W. Ryu, Food Technol. Biotechnol. 44 (2006) 163−172.
[149] M. Sauer, D. Porro, D. Mattanovich, P. Branduardi, Trends. Biotechnol. 26 (2008) 100−108.
[150] J. Lunt, Polym. Degrad. Stab. 59 (1998) 142−152.
[151] D. Carlotta, J. Polym. Environ. 9 (2002) 63−84.
[152] J.R. Dorgan, H. Lehermeier, M. Mang, J. Polym. Environ. 8 (2000) 1−9.
[153] N. Nagasawa, A. Ayako, S. Kanasawa, T. Yagi, H. Mitomo, F. Yoshii, et al., Nucl. Instr. Meth. Phys. Res. B 236 (2005) 611−616.
[154] F. Yang, R. Murugan, S. Wang, S. Ramakrishna, Biomaterials 26 (2005) 2603−2610.
[155] D.R. Kelsey, US Pat. US5705575 (1998).
[156] M. Arndt, DuPont creates carpet from corn, but should it. <http://www.businessweek.com/innovate/next/archives/2010/05/dupont_creates_carpet_from_corn_but_should_it.html>.
[157] SmartStrand carpet made with DuPont Sorona®, <http://shop.carpetmart.com/dupont-sorona-carpet>.
[158] M.V. Guettler, M.K. Jain, US Pat. US5521075 (1996).
[159] E. Delwiche, J. Bacteriol. 59 (1950) 439−442.

[160] Succinic acid applications, Thirumalai Chemicals. <https://www.thirumalaichemicals.com/images/sacapplications.pdf>.

[161] L. Stols, M.I. Donnelly, Appl. Environ. Microbiol. 63 (1997) 2695–2701.

[162] J.W. Lee, H.U. Kim, S. Choi, J. Yi, S.Y. Lee, Curr. Opin. Biotechnol. 22 (2011) 758–767.

[163] T. Tang, T. Moyori, A. Takasu, Isomerization–free polycondensations of cyclic anhydrides with diols and preparation of polyester gels containing cis or trans carbon double bonds via photo-croslinking and isomerization in the gels, Macromolecules 46 (2013) 5464–5472.

[164] J. Gowsika, R. Nanthini, Synthesis, characterization and in vitro anticancer evaluation of itaconic acid based random copolyesters, J. Chem. (2014). Available from: http://dx.doi.org/10.1155/2014/173814.

[165] A. Gandini, T.M. Lacerda, From monomers to polymers from renewable resources: recent advances, Prog. Polym. Sci. (2015). Available from: http://dx.doi.org/10.1016/j.progpolymsci.2014.11.002.

[166] T.J. Farmer, R.L. Castle, J.H. Clark, D.J. Macquarrie, Synthesis of unsaturated polyester resin from various bio-derived platform molecules, Int. J. Mol. Sci. 16 (2015) 14912–14932.

[167] S. Madbouly, C. Zhang, M.R. Kessler, Bio-Based Plant Oil Polymers and Composites, Elsevier, Oxford, UK, 2016.

[168] Susterra® 1, 3 Propanediol, High-performance ingredient for unsaturated polyester resins, DuPont Tate & Lyle BioProducts. <http://www.duponttateandlyle.com/susterra/unsaturated_polyester_resins>.

[169] R. Lewarchik, Bio-based resins for coatings. <http://knowledge.ulprospector.com/2343/pc-bio-based-resins-for-coatings/#_ftn2>.

[170] BioPreferred coating categories, BioPreferred, United States Department of Agriculture. <www.biopreferred.gov>.

[171] Industrial resins, <nzic.org.nz/ChemProcesses/polymers/10A.pdf>.

[172] Biobased resin technology: Helping define high performance, PCI Paint & Coatings Industry. <http://www.pcimag.com/articles/98699-bio-based-resin-technology-helping-define-high-performance>.

[173] Biobased chemicals: Greener paints improving coverage, ICIS Chemical Business. <http://www.icis.com/resources/news/2015/03/06/9866307/bio-based-chemicals>.

[174] R. Blaauw, Biobased resins, Wageningen UR. <https://www.wageningenur.nl/en/show/Biobased-resins.htm>.

6 Additives and Modifiers for Biopolymers

6.1 Introduction

Biopolymers are rarely used on their own to make biodegradable products. As commonly used in the plastics compounding industry, different additives and modifiers are generally blended with the virgin bioresin to obtain useful performance properties that are similar to those of conventional polymers. Benefits of using additives are as follows [1]:

1. To enhance melt strength, thermal stability, and melt lubricity of a biopolymer during processing.
2. To improve the performance of the biopolymers, especially impact strength, heat resistance, flame resistance, gas barrier properties, antifogging properties, etc.
3. To stabilize the biopolymer so that less degradation occurs at the elevated temperatures encountered during processing.

Most additives fall into one of the following categories:

- Modifiers: such as plasticizers, nucleating agents, clarifiers, impact modifiers, blowing agents, colorants, and coupling agents
- Stabilizers: such as antioxidants, heat and UV stabilizers, fire retardants, antistatic agents, and fungicides
- Processing aids: such as lubricants and compatibilizer

Additives are also used to modify the appearance or to reduce the cost of the final product. With the intention to improve functional characteristics or the addition of a new function, additives can be incorporated into biopolymer in any proportions during any stage of polymerization or processing. A favorable way of incorporating various additives into a biopolymer is by using masterbatches. A masterbatch provides a convenient way of handling small amounts of critical ingredients, like specialized additives in higher concentrations than those

Introduction to Bioplastics Engineering.
DOI: http://dx.doi.org/10.1016/B978-0-323-39396-6.00006-3

occurring in normal mixture for subsequent dilution. Additives for biopolymers must also not have any adverse environmental or health effect and not compromise the biodegradability of the material [2].

6.2 Types of Additives and Modifiers

Similar to conventional plastics, bioplastics with inherent weaknesses in processability and physical properties need help from additives. These weaknesses stand out especially in the current generation of biopolymers. Different products have different requirements of the package; however, using additives usually depends on the end-use application of the product. Some basic additives that are added to improve the properties and characteristics of the final product made from biopolymer include:

- Plasticizers, impact modifiers, colorants including dyes and pigments
- Antioxidants, light and UV stabilizers
- Antistatic, releasing, denesting, matting, and antiaging agents
- Fluorescent brighteners, dispersants, lubricants, thickeners, chain extenders, and coupling agents
- Cross-linking and nucleating agents
- Heat and viscosity stabilizers
- Antifogging, antistatic, foaming, mildew proof, antibacterial, and drying agents
- Flame retardants

In addition to the traditional polymer additives, the first-generation additives are being developed specifically to improve processing and properties of biopolymers that are being developed and introduced into the market. So far, manufacturers have only focused on developing additives for polylactic acid (PLA). This is because PLA is the most widely used biopolymer. However, more and more suppliers are now developing additives for a range of other bioresins. MHG biopolymers, formerly known as Danimer LLC and Cereplast, have started to offer compounds improved through their proprietary formulations and processing. These compounds can be added directly into the bioresin or an additive concentrate can be used. Twin-screw extrusion is typically used for dispersing additives in the concentrate [3–5].

Limited availability of additives for many biopolymers has posed challenges in the way they can be used. This is mainly because of two main reasons—(1) additive development for most biopolymers is still in the development phase and (2) the type of biopolymer (biodegradable vs bioderived and renewable types). In essence, additives for biopolymers can be screened into three types:

- Type 1 — These types of additives involve traditional additives that have no adverse effect on health or the environment and do not compromise resins compliance with compostability standards.
- Type 2 — These types of additives involve renewable additives derived from natural sources, but not necessarily biodegradable, for use in durable products.
- Type 3 — These types of involve additives that are both renewable and biodegradable, which are a good fit for single-use or short-lived products.

6.2.1 Type 1 Additives

These types of additives include minerals that have been used to modify properties in conventional plastics for decades. Minerals are inherently biocompatible and their addition to biopolymers can help reduce the carbon footprint of the final product. Examples of Type 1 additives are EMforce Bio additive, Polestar 200P, Eco-Delink, and CelTiX [6,7]. Some of the key properties offered by these mineral additives are as follows:

- Improve the dimensional stability
- Improve the melt strength
- Improve impact strength
- Increased opacity
- Increased stiffness
- Improved processability

6.2.2 Type 2 Additives

These types of additives are derived from natural sources, but are not necessarily biodegradable. Their typical applications are in durable goods. Examples of Type 2 additives are PolyOne's OnCap BIO Antifog, Arkema's Biostrength, DuPont's Biomax Strong, and DuPont's

Elvaloy. Some of the key properties offered by these additives are as follows:

- Increased processability
- Improve melt stability
- Improve toughness
- Reduce brittleness
- Retain transparency
- Enhance heat resistance
- Increase heat distortion temperature (HDT)

6.2.3 Type 3 Additives

These types of additives are derived from natural resources and are biodegradable. These additives are fit for single-use or short-lived products. Example of Type 3 additives are Clariant's CESA-natur product line and Clariant's RENOL-natur color product line. These new masterbatches from Clariant—including slip/antiblock agents, UV/light stabilizers, antioxidants and antistatic agents feature not only a biopolymer carrier resin but also naturally occurring substances that are completely renewable, biodegradable and, in most cases, compostable.

6.3 Additives and Modifiers for Bioolefins

Polyolefins resist all kinds of biodegradation even when they are loaded with an additive that induces polymer chain breakage under certain conditions (oxygen, sunlight, etc.). A polyolefin product might require years to fully degrade especially if it still contains antioxidants that interfere with degradation or it is buried in a landfill. However, there are number of suppliers who promote polyolefin additives and degradation after disposal.

Fig. 6.1 shows highly transparent products made when PP clarifier, Millad NX8000 was added with the resin [8]. Milliken's latest PP clarifier, Millad NX8000 is now able to provide clarity in thicker sections setting up new standards for PP clarity. In comparison with Milliken's previous clarifier, Millad 3988i, Millad NX8000 produces 50% less haze at any part thickness. It is also known that Millad NX8000 melts at a lower temperature than Millad 3988i, allowing easier compounding. The suggested loading ranges from 0.4% to 0.5% and can be adjusted depending on the required transparency [8]. Fig. 6.2 shows the

Figure 6.1 Milliken's latest PP clarifier, Millad NX8000 [8].

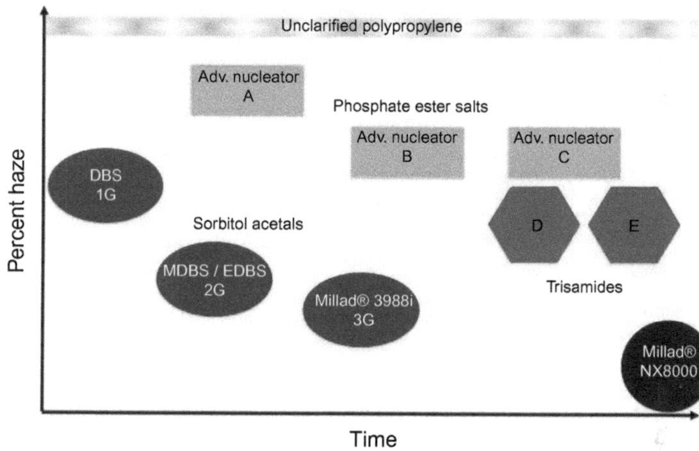

Figure 6.2 Evolution of Millad NX8000 from the range of Milliken's polypropylene clarifying agents [9].

evolution of Millad NX8000 from the range of Milliken's polypropylene clarifying agents [9].

Milliken's latest nucleant, Hyperform HPN-20E provides unique polypropylene (PP) physical properties that combine balanced bidirectional stiffness and excellent impact performance. Some of Hyperform examples are in automotive, appliances, pipe extrusion, closures, and industrial and are shown in Fig. 6.3. Some of the key properties of Hyperform HPN-20E are as follows [10]:

- Excellent stiffness/impact balance
- Unique shrinkage for dimensional control
- Superior coefficient of linear thermal expansion and HDT performance
- Excellent nucleation performance to maximize processing speed

Figure 6.3 Typical applications where Hyperform HPN-20E is used [10].

Fig. 6.4a–d shows comparison of Hyperform HPN-20E with clarifiers such as acetal sorbitols and nucleators such as talc, sodium benzoate, and phosphor ester salt for a range of properties such as shrinkage characteristics, coefficient of thermal expansion, bidirectional stiffness, and charpy notched impact. As shown in Fig. 6.4a, Hyperform HPN-20E has good bidirectional dimensional stability in comparison to other additives. Percent shrinkage of HPN-20E is around 1.27 in machine direction while 1.33 in transverse direction [10].

Fig. 6.4b shows comparison of coefficient of thermal expansion of Hyperform HPN-20E with other additives. Although, Hyperform expands more in machine direction than in transverse direction, it still has the lowest thermal expansion among other clarifiers and nucleators. Fig. 6.4c shows comparison of bidirectional stiffness of Hyperform HPN-20E with other additives. The flexural modulus of HPN-20E is 1800 MPa which is similar to additives with nucleating agents [10]. Hyperform HPN-20E has excellent impact properties with the charpy izod impact value of 3.2 kJ/m^2 at room temperature [10].

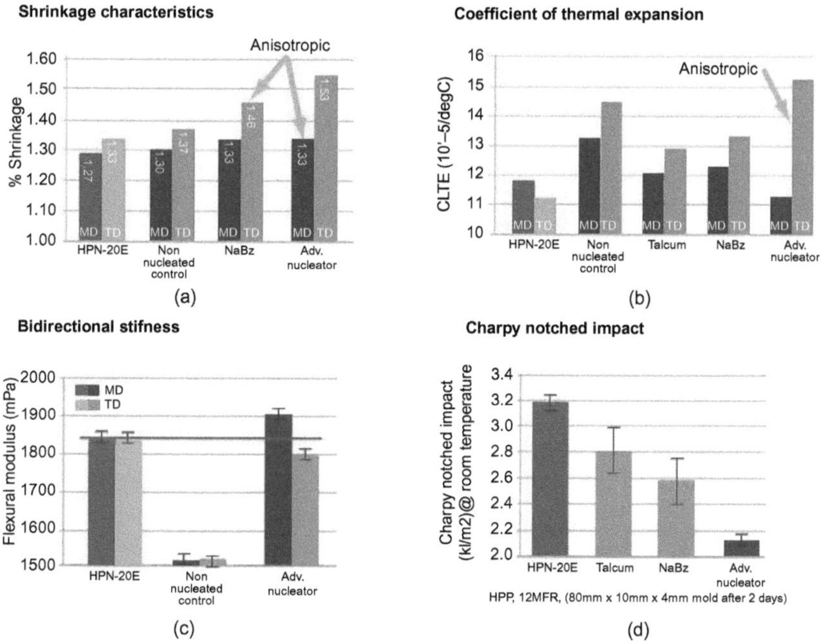

Figure 6.4 Comparison of Hyperform HPN-20E with other clarifiers and nucleators for (a) shrinkage characteristics, (b) coefficient of thermal expansion, (c) bidirectional stiffness, and (d) charpy notched impact [9,10].

IRGACLEAR XT 386 from Ciba Speciality Chemicals is a high-efficiency, next-generation clarifier for polypropylene. It is based on new trisamide chemistry which matches the clarity of Millad 3988i but at-use levels is around 150 ppm [11]. Extremely effective at low concentrations, it enables see-through clarity of polypropylene products. IRGACLEAR XT 386 has been derived from a new chemistry platform for clarifiers, which features high thermal stability and high chemical integrity [11]. Fig. 6.5 shows a difference in clarity in a product when used without and with IRGACLEAR XT 386.

In addition, IRGACLEAR XT 386 tolerates higher compounding and processing temperatures, compared with many other clarifiers, and maintains its clarifying power even when subjected to multiple processing cycles. The high stability of IRGACLEAR XT 386 allows for output improvements and higher quality polypropylene products [12].

Ciba's IRGASURF SR 100 is an antiscratch additive for PP and TPO. It is a polymeric material added at 1–3% to boost scratch resistance in auto exterior and interior parts [13]. IRGASURF SR 100 has an excellent scratch performance up to 15 N on the cross-cut Erichsen

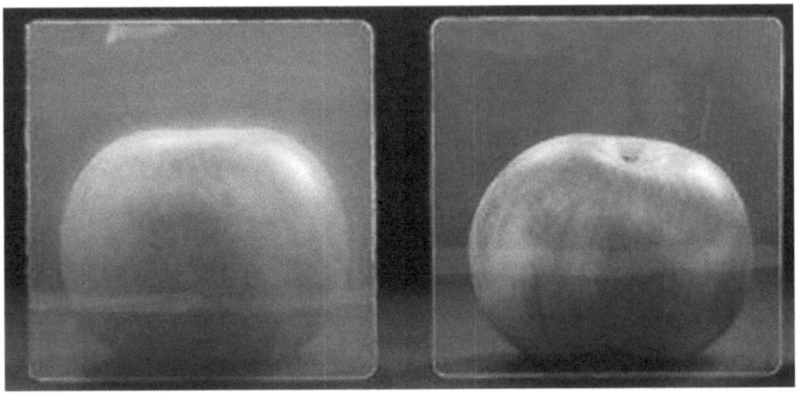

Figure 6.5 Left: without IRGACLEAR XT 386, and right: with IRGACLEAR XT 386 [11].

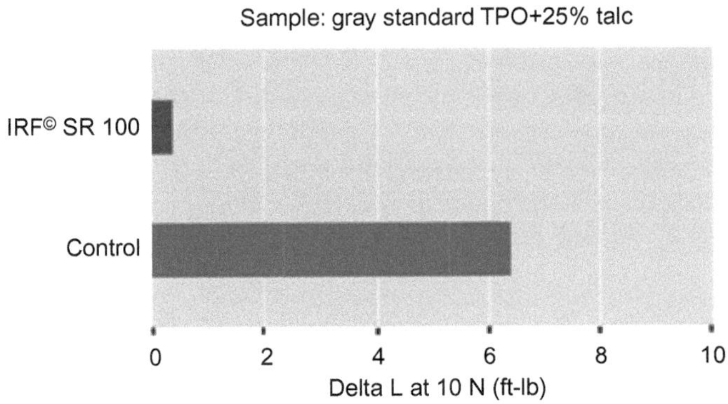

Figure 6.6 The Erichsen Cross-Cut Scratch Test comparing scratch visibility at 10 N scratch load on N127 grain plaques [13].

tests as shown in Fig. 6.6. It also performs efficiently in combination with various light stabilizers and does not show any detrimental surface effects such as stickiness or exudation during processing and use [14].

6.4 Additives and Modifiers for Biopolymers

6.4.1 Toughness

DuPont's Biomax Strong 100 and 120 improve toughness and reduce brittleness in PLA rigid molded and thermoformed parts. At 1–3% loadings, they reportedly outperform competitive tougheners with little effect on transparency. Cast sheets containing Biomax Strong exhibit

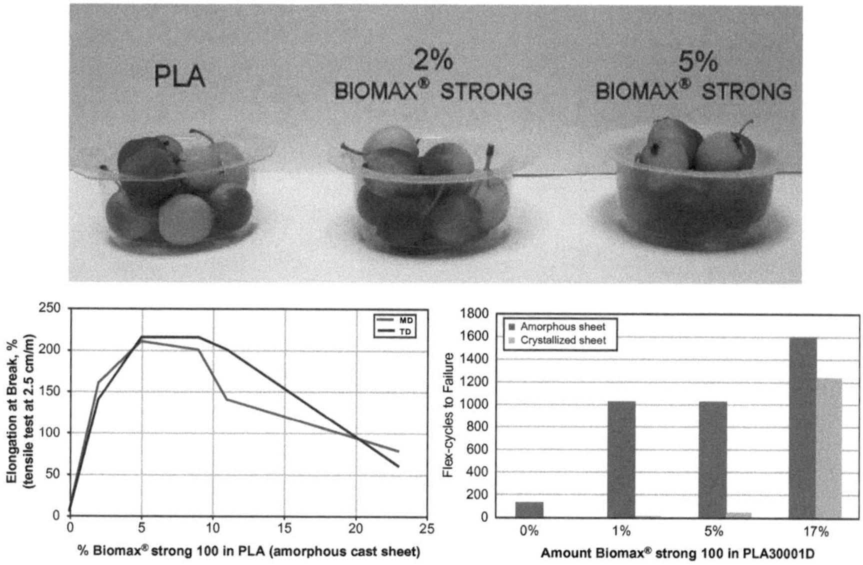

Figure 6.7 Effect of Biomax Strong loading levels on mechanical properties [15].

improved cutting and trimming can withstand repeated flexing and exhibit increased elongation at break. Biomax Strong also acts as a processing aid by stabilizing PLA viscosity against thermal degradation and reducing torque. The additive marginally reduces clarity and can be dosed directly during processing. Fig. 6.7 shows the effect of Biomax Strong loading levels on the mechanical properties of the articles made with it. Biomax Strong is not biodegradable, but at low loading levels it does not affect the compostability of articles made with it [15].

Arkema offers new Biostrength 130 acrylic modifier, which retains adequate transparency for translucent PLA applications [2]. Biostrength 150 impact modifier is a high-efficiency modifier that increases the impact strength of PLA. It is recommended for opaque applications or applications that do not require high transparency. It is typically used in sheet extrusion, calendering, injection molding, and thermoforming applications with typical loading range from 2% to 6% by weight. Fig 6.8 shows exponential increase in impact properties with increase in weight percent impact modifier [16].

Fig. 6.9 shows the effect of Biostrength 150 impact modifier on impact properties at low temperature on PLA. At low loading levels, the impact properties of PLA are very similar at room temperature, $-20°C$ and $-30°C$. However, at higher loading levels (~6 wt%), a noticeable difference in impact properties at three different

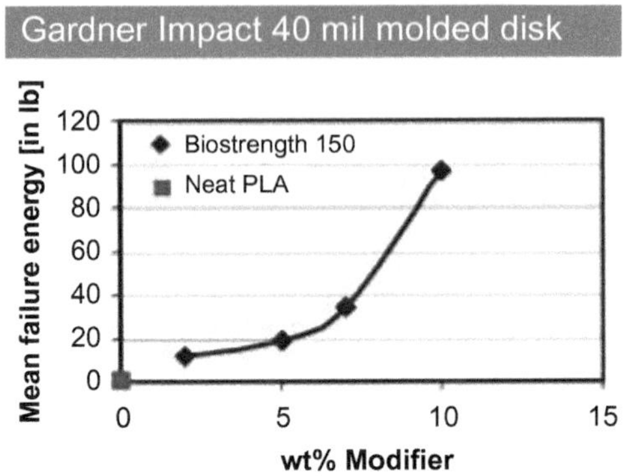

Figure 6.8 Effect of impact modifier on impact properties of PLA [16].

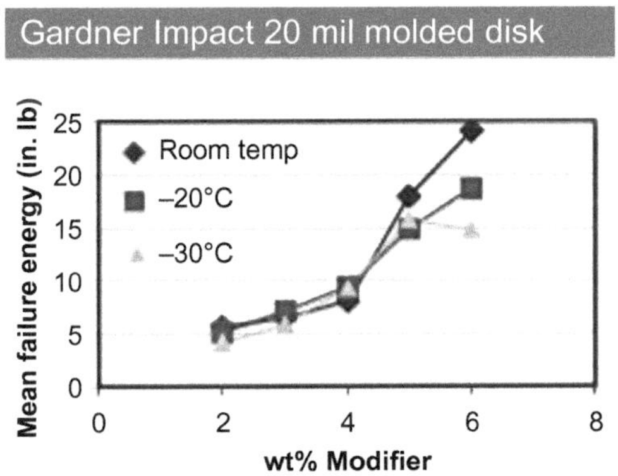

Figure 6.9 Effect of impact modifier on impact properties at low temperatures on PLA [16].

temperatures can be seen. Blends of Biostrength 150 impact modifier and PLA can be processed in the same equipment and under the same processing conditions recommended for PLA [16].

In applications which require high toughness and transparency, Arkema's Biostrength 280 impact modifier is recommended. Typical use levels range from 2% to 6% by weight. Blends of Biostrength 280 impact modifier with PLA can be processed in the same equipment and under the same processing conditions recommended for PLA. Fig. 6.10 shows the effect of Biostrength 280 impact modifier on the impact strength of PLA. The impact strength is seen to increase with increase in weight percent loading of Biostrength 280. However, at approximately 10 wt% loading levels, the impact strength starts to level off and does not see a significant increase in the impact strength [17].

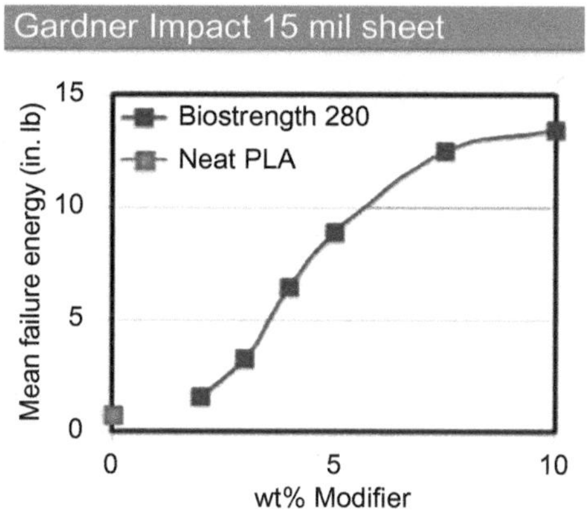

Figure 6.10 Effect of Biostrength 280 concentration on the impact strength of PLA [17].

Arkema's Biostrength Advantage is another composite acrylic core–shell impact modifier that provides superior impact performance in PLA at low loading levels. Advantage impact modifier with PLA can be processed in the same equipment and under the same processing conditions recommended for PLA [18].

6.4.2 Processability

One of the most common limitations of biopolymers especially PLA is their low melt strength. This low melt strength does not allow processing using extrusion, blow molding, foaming, and deep-draw thermoforming [19]. When Arkema's Biostrength 700 acrylic copolymer is added at a loading level of 1–5%, it improves melt strength and enhances processability of PLA. Fig. 6.11 shows improvement in melt strength of PLA by adding 4 wt% Biostrength 700. It is transparent in PLA and decreases sagging and necking phenomenon of the PLA melt, providing greater ease in processing and higher quality finished products [19].

In addition, Biomax Strong 100 and 120 ethylene copolymers, developed to toughen PLA, also act as processing aids that significantly reduce screw torque and improve melt stability. Some of the attributes of Biomax Strong 100 and 120 in PLA are shown in Table 6.1 [15].

While PLA has adequate melt strength for sheet, its relatively low melt strength limits its capability in processes that require higher melt strength, such as blown film extrusion and foaming. PLA has inherently low melt viscosity, and is also vulnerable to thermal, oxidative,

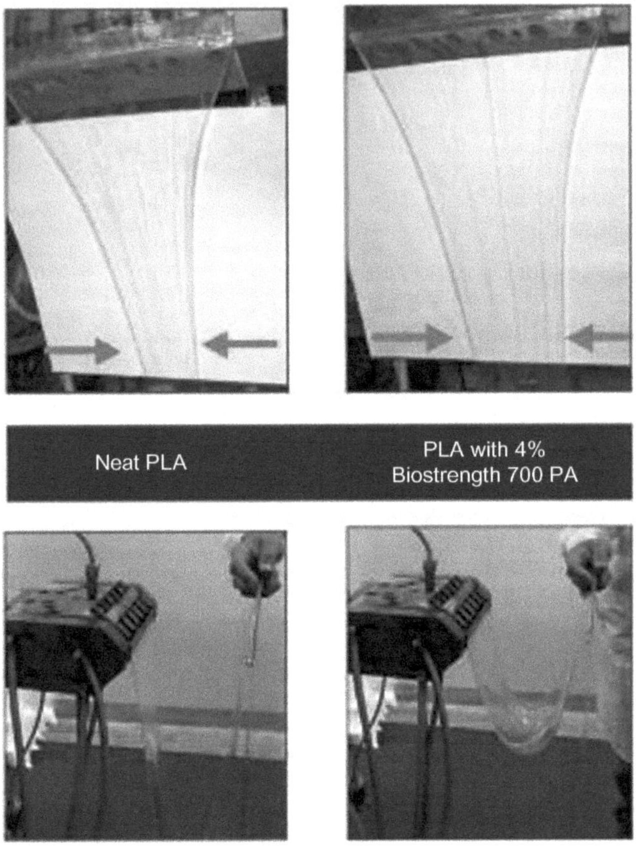

Figure 6.11 PLA with 4% Biostrength 700 increases melt strength of PLA [3,19].

Table 6.1 Attributes of Biomax Strong 100 and 120 [15]

Biomax Strong 100 and 120
• Special chemistry for PLA • Rapid melt-dispersion • Pelletized, efficient extruder feeding • Blends containing 5 wt% or less modifier maintain contact clarity similar to clarified PP while blends containing 5–15 wt% modifier maintain opacity similar to unclarified PP

and hydrolytic degradation, which causes chain scission and a further loss of molecular weight and lower viscosity. Clariant's CESA-Extend reconnects the polymer chains and creates a branched network that improves melt strength and increases tensile strength properties.

Figure 6.12 PLA foam tray [3]. Photo: Clariant Masterbatches.

It also repairs the damage and restores the resin's intrinsic viscosity [20,21]. Fig. 6.12 shows Clariant's CESA-extend used in packaging applications [3].

Clariant also offers masterbatches using conventional synthetic additives in a PLA carrier, including CESA-bloc antiblock masterbatches and CESA-light UV stabilizer masterbatches. Among the products being evaluated are slip additive masterbatches, based on naturally occurring waxes that can produce a coefficient of friction in PLA films similar to that achieved with synthetic waxes [21].

6.4.3 Heat Resistance

Heat distortion temperature or heat deflection temperature (HDT) is the temperature at which polymer deforms under a load. It is used to determine short-term heat resistance. There are biopolymers that lose rigidity over a narrow temperature range. Heat deflection of biopolymer can be improved by using HDT modifiers and allow biopolymer to withstand higher processing and use of temperature. For example, PLA has a low HDT of around 50–60°C which causes PLA products to deform during storage and PLA pellets to stick during transportation and handling [22].

There are many big players in the market that make HDT modifiers for different biopolymers. MHG Biopolymers, formerly known as Danimer LLC, provides a renewable additive that improves the heat deflection of PLA to 88°C. This improvement in the heat deflection of PLA results in better performance characteristics of products. Products

Figure 6.13 Product made using PLA using MHG additive [23].

range from thermoformed parts such as packaging tray to injection molded parts such as plastic cup, spoon, and plates as shown in Fig. 6.13 [22–24]. This material is not transparent, and will produce some increased opacity, depending upon the percent additive used.

NatureWorks reports that the heat resistance of PLA can be improved by adding cellulose fibers to biopolymer matrix. Similarly, when compatibilized starch particles are added to PLA, it reduces the cost and enhances the modulus and heat resistance. Additives such as nucleating agents can also give slight increases in heat resistance [25].

DuPont Biomax Thermal 300 is an ethylene copolymer with a nucleating agent designed to modify PLA for improved thermal and toughness properties in packaging and industrial applications. PLA thermoformed packages can be stored shelf-stable, shipped normally, and even reheated in the microwave without deformation [26]. It is recommended that 5 wt% or less of Biomax Thermal 300 should be blended with PLA in order to maintain clarity of the product. Fig. 6.14a and b shows comparison of PLA tray with 2 wt% Biomax Thermal 300 exposed to 70°C for 1 h with PLA tray without Biomax Thermal 300 [27]. The PLA tray which had no Biomax Thermal 300 degrades on exposure to heat in comparison to the PLA tray which contained the additive.

Another manufacturer of additives, PolyOne is always exploring technologies to improve heat resistance of biopolymers such as PLA and starch blends. They have developed polymer blends containing PLA that can raise the HDT from around 70°C to around 100°C [3].

(a) (b)

Figure 6.14 (a) PLA tray with 2% Biomax Thermal 300 exposed to 70°C for 1 h, and (b) PLA tray without modifier exposed to 70°C for 1 h. [27].

However, the non-PLA blend components are neither bioderived nor biodegradable, and the system is opaque [2,22]. PolyOne is also working with polyhydroxybutyrate valerate (PHBV) resins that offer improved heat resistance [28].

6.4.4 Foamability

Clariant's HYDROCEROL chemical foaming agents creates a finer, more durable foam-cell structure [29]. It is used in PLA foam extrusion and increasingly as a nucleating agent in direct-gas foaming to produce the uniform, fine-celled structure. Adding CESA-extend to the foam, in combination with HYDROCEROL, has been shown to enhance molecular weight and melt-strength characteristics resulting in a more easily controllable process. The HYDROCEROL masterbatches can achieve a highly consistent structure of cells as small as 60 μm in diameter. It is actively used in injection molding, extrusion, and thermoforming processes similar to the products shown in Fig. 6.15 [30].

Another manufacturer, Reedy International, has introduced two new foaming agents designed for PLA. Safoam RPC-20MS1 and 20MS2 is an endothermic foaming agent with built-in melt enhancing properties. They perform best when used in low-density PLA applications such as meat trays, and has potential for medium- to low-density foamed PLA deli, margarine, or ice-cream containers [2,31].

6.4.5 Mechanical Properties

Tianan Biologic, Metabolix, and Meredian are the key biopolymer suppliers of the virgin PLA resins. The virgin polymer is difficult to

Figure 6.15 Some of the applications of Clariant's HYDROCEROL [30].

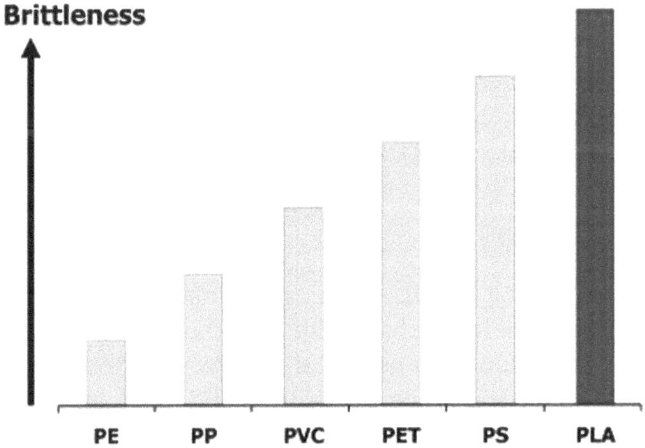

Figure 6.16 Polylactic acid (PLA) is more brittle than any polymer [32].

process because of its inherent properties. However, there has been emphasis on the importance of polymer modification by using additives. As shown in Fig. 6.16, brittleness is a key inherent property of PLA. This has led NatureWorks to study the importance of impact modification [32]. MHG Biopolymers have also used bio-based additives to improve mechanical properties of PLA [23].

BASF Ecovio resins are a combination of PLA with Ecoflex, petroleum-based, biodegradable, aliphatic–aromatic copolyesters [33]. They are used typically in packaging applications such as foamed clamshells as shown in Fig. 6.17.

Specialty Mineral's EMforce Bio is a high aspect ratio, carbonate-based engineered mineral, and fully compostable designed to improve the impact toughness of amorphous and semicrystalline PLA for opaque applications. When dispersed in the polymer matrix, the mineral

Figure 6.17 BASF's new biodegradable Ecovio blend for foamed clamshells [33].

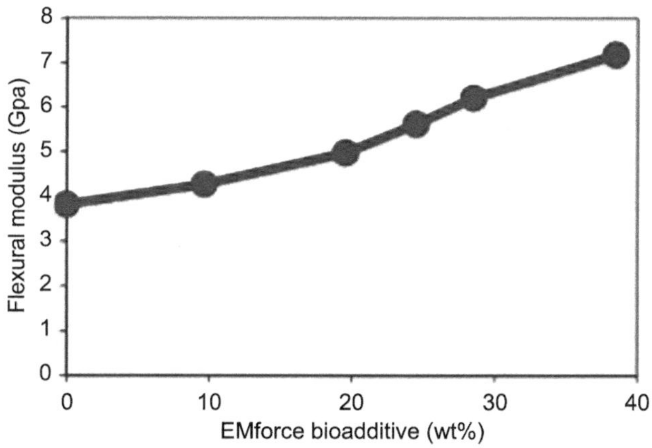

Figure 6.18 Flexural modulus of PLA with EMforce Bio [3].

additive absorbs energy and slows crack propagation, thus toughening the biopolymer. Special Mineral recommends 20–40 wt% EMforce Bio to drastically improve toughness, impact resistance, and flexural modulus as shown in Fig. 6.18 [3].

Some of its other key benefits are as follows:

- Change polymer failure mode to Brittle to 100% Ductile state
- Greatly improve impact strength
- Improved tear and puncture resistance

Figure 6.19 Impact modifiers and processing aids for PLA comprise Arkema's new Biostrength family of additives for biopolymers [34].

- Achieve higher color of composites
- Extension of resin for cost savings
- Nonsilica-based mineral

Speciality Mineral's EMforce Bio is also used as a nucleating agent; however, it results in higher crystallinity. Arkema's Biostrength is a core–shell impact modifier, which when added to PLA improves the impact strength of PLA. Typical loading levels are in the range of 4 to 7 wt% [34]. It is typically used in packaging applications as shown in Fig. 6.19 [34].

PolyOne has a range of impact modifier masterbatches, including those in its new OnCap Bio line for opaque and transparent biopolymer as shown in Fig. 6.20 [35]. The OnCap Bio additive concentrate provides improvement in the following properties to biodegradable resins:

- Antiblock
- Antistatic
- Denesting
- Slip
- UV protection
- Blue tone
- Antifog

Originally introduced by Rohm and Haas and now product of Dow Chemicals, Paraloid BPM-500 is an acrylic impact modifier which is used to improve impact in PLA and other biopolymers. Significant impact strength are obtained with levels as low as 3 wt%. Fig. 6.21 shows the impact performance of Paraloid BPM-500. A threefold

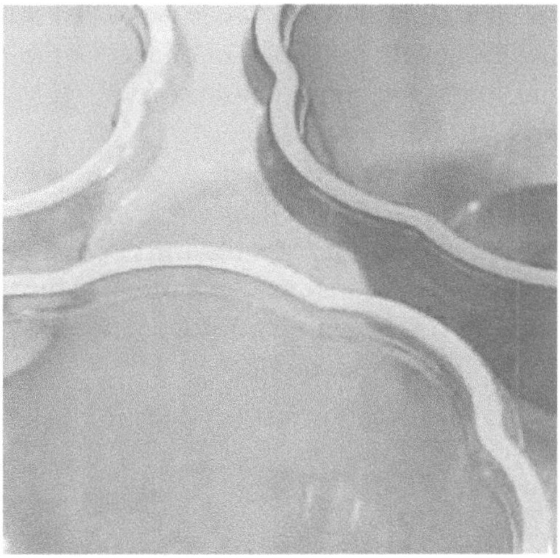

Figure 6.20 PolyOne OnCap bioadditives [35].

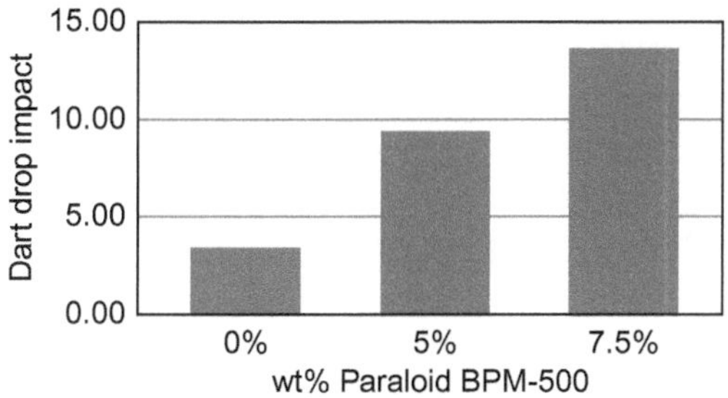

Figure 6.21 Impact performance of Paraloid BPM-500 [36].

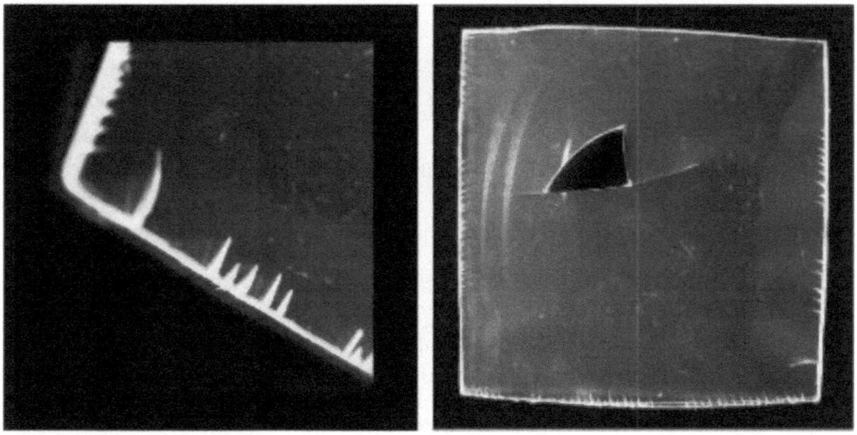

Figure 6.22 Brittle web fracture (left) and brittle fracture of a PLA container with Sukano PLA im S550 [32].

increase in the impact strength of PLA is observed when BPM-500 is used at 5 wt% loading levels [36]. It also enhances the resistance of the film to tearing and improves cutting. Addition of BPM-500 has a minimal effect on clarity of the PLA films.

Fig. 6.22 shows a brittle fracture in PLA container with Sukano PLA im S550. Sukano's patented, FDA-approved impact modifier concentrate, Sukano PLA im S550, improves the impact characteristics of PLA without reducing clarity, heat stability, or compostability [32,37]. Using biopolymer is essential for obtaining homogeneity and clarity. This also guarantees compostability of the product.

6.5 Effect on Properties

Most plastic additives are introduced into a polymer to enhance a specific property, whether to have ease in processability or to improve mechanical properties. The effect of a given amount of a given additive on polymer properties will depend on the additive physical properties and its chemical composition. Besides mechanical properties, additives also contribute to change in optical and electrical properties, chemical and weathering resistance, flammability and density of polymers among other properties. This section will discuss effect of additive on:

- Toughness
- Processability
- Heat resistance
- Foamability
- Mechanical properties

6.5.1 Toughness

Toughness is the ability of a material to absorb energy and deform without fracture. It is an important parameter that dictates whether or not a biopolymer can be used for an application which requires such properties. Impact modifiers generally increase the viscosity of the polymer. It also affects the number of other performance properties of the polymer, such as its optical properties, weatherability, processability, mechanical properties, flammability, and heat resistance apart from increasing toughness.

Impact modifiers provide a boost to the impact resistance of biopolymers. They can often be used to help correct the detrimental effects of other additives on the overall impact resistance. Impact modifiers improve the impact resistance of polymer by facilitating the ability of the polymer matrix to craze.

Impact modifiers have varying influences on melt rheology. During extrusion without appropriate adjustments of lubricants they lead to over- or underfused extrudate, which will have poor impact. Some of the impact modifiers may require a different degree of fusion in order to attain optimum effect [38].

6.5.2 Processability

Typically, processing aids are used when processing a biopolymer which lacks melt strength. As an example, PLA has poor melt strength and decreases with molecular weight. This makes PLA tough to process and therefore processing aids are used in processes such as extrusion and injection molding to reduce friction between the biopolymer melt and the metal, which facilitates processing. In addition, melt strength enhancers are also added to improve the processing of biopolymers [22]. Some other additives that are used to enhance polymer processability include:

- Heat stabilizer
- Lubricants
- Processing aids

Achieving a balance of these additives is important in obtaining adequate processability and heat stability, satisfactory properties, and an acceptable level of performance. In general, additives containing an ester, an epoxy, or a diepoxy-polyether groups are able to enhance the processability of PLA [39].

6.5.3 Heat Resistance

One of the major shortcomings of biopolymers is their lack of resistance to deformation at higher temperatures. HDT is often used to determine short-term heat resistance of materials and is the temperature at which a test sample deflects under a constant load.

Improvement of thermal resistance of a biopolymer depends on the polymer structure. To increase the HDT of semicrystalline, polymers can be achieved by increasing the crystallinity of the polymer by using nucleating agents at optimum levels. This helps in increasing the HDT nearer to the melting temperature of the polymer. PLA is also blended with other polymers including other biopolymers such as PHBV which can improve PLA's heat resistance [37].

6.5.4 Foamability

Polymer foamability is evaluated by three measured parameters: the foaming temperature window, the correlation factor of foam density with foaming temperature, and blowing agents solubility [40]. While blowing agent solubility determines the extent of plasticization and the final density of the foam, the foaming temperature window determines the lower and upper bound foaming temperatures at a given foaming condition [41].

To have a useful process, the processing window is shown in Fig. 6.23, in which temperature and viscosity characteristics are bounded by the degree of expansion and cell integrity. As noted from Fig. 6.23, amorphous biopolymers have a broad foam window compared to crystalline biopolymers [41,42]. This means that amorphous biopolymers can be processed at significantly higher temperatures than crystalline biopolymers before they start to foam.

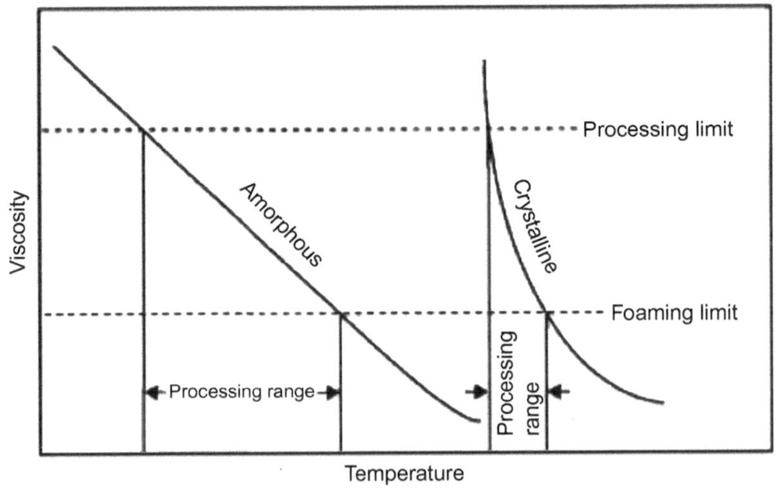

Figure 6.23 Foam processing window [41,42].

6.5.5 Mechanical Properties

Poor mechanical properties of virgin biopolymers are a point of concern. By chemically modifying the polymer, introducing additional reactive groups and by adding additives, mechanical properties of a polymer can be achieved. Mechanical properties also depend on biopolymers molecular weight, crystallinity, and orientation.

Fillers, usually in particulate or fibrous form, are blended together with the polymer. These additives alter the molecular structure of the polymer or add a second phase to the polymer affecting mechanical properties [43]. Composites from biopolymer provide a route to strong materials without reducing their ability to biodegrade. By combining cross-linking and fiber additives, protein-based polymers have observed mechanical properties similar to polystyrene [44].

Blending with plasticizers improve biopolymers that are brittle with an elongation range of 2–5 wt%. A plasticizer is commonly used to improve the processability, flexibility, and stretchability of a polymer because it can reduce the intermolecular force and increase the intermolecular mobility [45]. In some cases, improvement in tensile strength, and modulus can be achieved using a maximum of 20 wt% plasticizer [46].

References

[1] M. Niaounakis, Biopolymers: Processing and Products, Elsevier, Waltham, 2015.

[2] E. Miller, Introduction to Plastics and Composites: Mechanical Properties and Engineering Applications, Marcel Dekker Inc., New York, 1996.

[3] L.M. Sherman, Enhancing biopolymers: additives are needed for toughness, heat resistance and processability, Plast. Technol. 54 (7) (2008) 58–63. <http://www.ptonline.com/articles/enhancing-biopolymers-additives-are-needed-for-toughness-heat-resistance-processability>.

[4] MHG Biopolymers. <www.mhgbio.com>.

[5] Cereplast. <www.cereplast.com>.

[6] EMforce® Bio additive, Speciality Minerals. <http://www.mineralstech.com/Documents/MTI/SDS%20%26%20Technical%20Data%20Sheets/Adams/Adams%20Additive%20Products/EMforce%20Bio.pdf>.

[7] Imerys, Mineral Additives for Biopolymers and Biodegradable Polymers. <http://www.imerys-perfmins.com/pdf/FFAE-P005_Biopolymer_Brochure_A4.pdf>.

[8] Millad® NX™ 8000 for Thermoforming. <http://millikenchemical.com/millad-nx-8000-thermoforming/>.

[9] D.Lukaszewski, T.Hauck, Polypropylene clarification and nucleation, in: Moscow International Conference PP 2008. <http://www.creonenergy.ru/upload/iblock/072/Polypropylene%20Clarification%20s%20Nucleation.%20Dariusz%20Lukaszewski,%20Milliken%20Europe.pdf>, 2008.

[10] Milliken® Hyperform® HPN-20E Nucleating Agent for Polypropylene. <http://millikenchemical.com/site/user/files/1/Hyperform-HPN-20E-Nucleating-Agent-Polypropylene.pdf>.

[11] Clarifier for Polypropylene Approved by FDA, Flexible Packaging. <http://www.packagingdigest.com/flexible-packaging/clarifier-polypropylene-approved-fda>, October 2008.

[12] Ciba Speciality Chemicals Introduces Next Generation Clarifier for Polypropylene. <http://polymer-additives.specialchem.com/news/industry-news/ciba-specialty-chemicals-introduces-next-generation-clarifier-for-polypropylene>.

[13] Irgasurf® SR 100 B, Technical Information. <http://www.telko.com/files/images/telko/ru/basf/antiscratch/irgasurf_sr_100_b_tds.pdf>.

[14] Irgasurf® SR 100, Technical Information. <http://www.plas2006.com/UploadFile/TopicFile/20063171993352.doc>.

[15] DuPont™ Biomax Strong®. <http://www.dupont.com/content/dam/assets/industries/packaging-printing/pdf/Biomax_Strong.pdf>.

[16] Arkema Biostrength® 150, Product Data Sheet. <http://www.additives-arkema.com/export/sites/acrylicmodifiers/.content/medias/downloads/literature/biostrength-150-impact-modifier.pdf>.

[17] Arkema Biostrength® 280, Product Data Sheet. <http://www.additives-arkema.com/export/sites/acrylicmodifiers/.content/medias/downloads/literature/biostrength-280-impact-modifier.pdf>.

[18] Arkema Biostrength® Advantage, Technical Information. <http://www.additives-arkema.com/export/sites/acrylicmodifiers/.content/medias/downloads/literature/biostrength-advantage.pdf>.

[19] Arkema Biostrength 700, Technical Information. <http://www.additives-arkema.com/export/sites/acrylicmodifiers/.content/medias/downloads/literature/biostrength-700-pa.pdf>.

[20] CESA®-extend Processing Improver, Clariant Products and Solutions. <http://www.clariant.com/Solutions/Products/2014/08/12/11/12/CESA-extend>.

[21] R.A. Auras, L.-T. Lim, S.E.M. Selke, H. Tsuji, Poly(lactic acid): Synthesis, Structures, Properties, Processing, and Application, Wiley Publishers, Hoboken, NJ, 2010.

[22] M. Niaounakis, Biopolymers: Reuse, Recycling, and Disposal, Elsevier, Oxford, UK, 2013, p. 47.

[23] MHG Biopolymer—Additives Improve the Properties of Feedstock. <http://www.mhgbio.com/mhg-products/mhg-products-biopolymer-additives/>.

[24] MHG Products—Biopolymer Development and Production. <http://www.mhgbio.com/mhg-products/>.

[25] Technology Focus Report: Polylactic Acid Containing Fillers and Fibers. <http://www.natureworksllc.com/~/media/Technical_Resources/Properties_Documents/PropertiesDocument_Fillers-and-Fibers_pdf.pdf?la=en>.

[26] DuPont Packaging and Industrial Polymers, Biomax® Thermal Resins Product Data Sheet. <http://www.dupont.com/content/dam/assets/products-and-services/packaging-materials-solutions/assets/biomax_thermal_300.pdf>.

[27] DuPont Advances Sustainable Packaging With New Renewably Sourced Offering and Industry Program. <http://www.prweb.com/releases/2008/09/prweb1308624.htm>.

[28] PHBV-Based Biopolymer Compounds, A Case Study. <http://www.polyone.com/files/resources//PHBV-Based_Biopolymer_-_Consumer_Case_Study.pdf>.

[29] Natural, Renewable Additive Masterbatches Expand Clariant's Green Portfolio. <http://omnexus.specialchem.com/news/industry-news/natural-renewable-additive-masterbatches-expand-clariant-s-green-portfolio>.

[30] Lighten Up: HYDROCEROL® Chemical Foaming and Nucleating Agents. <http://www.clariant.com/~/media/Files/Business-Units/Masterbatches/Masterbatches-_-Products/Hydrocerol_Chemical_Foaming_and_Nucleating_Agents_Flyer_en.pdf>.

[31] Reedy International. <http://www.reedyintl.com/products/safoam.php>.

[32] J. Markarian, Biopolymers present new market opportunities for additives in packaging, Plast. Addit. Compd. 10 (3) (2008) 22–25, Elsevier.

[33] J.H. Schut, Foamed PLA shows promise in biodegradable meat trays, Plast. Technol. 53 (12) (2007) 39–41. <http://www.ptonline.com/articles/foamed-pla-shows-promise-in-biodegradable-meat-trays>.

[34] M. Naitove, Additives and colorants score dramatic advances, Plast. Technol. 54 (3) (2008) 66–71. <http://www.ptonline.com/articles/additives-and-colorants-score-dramatic-advances>.

[35] OnCap™ BIO Additives, Product Data Sheet, PolyOne. <http://www.polyone.com/en-us/products/bioproducts/Pages/OnCapBIOAdditives.aspx>.

[36] PARALOID™ BPM-500 Acrylic Impact Modifier for Polylactic Acid Resins, Rohm Haas, Product Data Sheet. <http://www.dow.com/assets/attachments/business/plastics_additives/paraloid_bpm/paraloid_bpm-500/tds/paraloid_bpm-500.pdf>.

[37] Materials Additives and Blends Take PLA Upscale, Plast. Technol. <http://www.ptonline.com/articles/materials-additives-blends-take-pla-upscale>, 2010.

[38] M.N. Subramanian, Plastics Additives and Testing, Wiley Publishers, Salem, MA, 2013 (Chapter 3).
[39] S. Thomas, P.M. Visakh, Handbook of Engineering and Speciality Thermoplastics, vol. 3, Wiley Publishers, Salem, MA, 2011, p. 358.
[40] J. Xu, Microcellular Injection Molding, Wiley Publishers, Hoboken, NJ, 2011 (Chapter 4).
[41] M.N. Subramanian, Plastics Additives and Testing, Wiley Publishers, Salem, MA, 2013 (Chapter 8).
[42] S.T. Lee, Nucleation in Foam Extrusion. J. Cell. Plast., 37 (2001) 221–230 (Sage Publications).
[43] M.P. Groover, Fundamentals of Modern Manufacturing: Materials, Processes, and Systems, fourth ed., Wiley Publishers, Hoboken, NJ, 2010 (Chapter 8).
[44] E.S. Stevens, Green Plastics: An Introduction to the New Science of Biodegradable Plastics, Princeton University Press, Woodstock, UK, 2002.
[45] R. Wool, X.S. Sun, Bio-based Polymers and Composites, Academic Press, London, UK, 2005.
[46] V.K. Thakur, M.K. Thakur, Eco-friendly Polymer Nanocomposites: Chemistry and Applications, Springer, New Delhi, India, 2015.

7 Processing Biodegradable Polymers

7.1 Introduction

Biodegradable polymers like any other polymer can be processed using conventional techniques such as injection molding, extrusion, and compression molding. In addition to polymer processing, special attention needs to be given to polymer. For example, moisture needs to be removed from the polymer prior to processing to prevent hydrolytic degradation of polymer. Polymers should be dried for a predetermined time to remove moisture. In addition, the processing should be performed under controlled humidity [1].

Most biodegradable polymers are synthesized by ring-opening polymerization reaction. During injection molding and extrusion process, excessively high processing temperatures may result in monomer formation due to the thermodynamic equilibrium which exists between the forward and the reverse reaction. This excess monomer can act as a plasticizer affecting material mechanical properties and degradation kinetics. Therefore, these materials should be processed at the lowest temperatures. Biodegradable polymers are distinctly different than regular polymers in material characteristics. The distinct narrow processing window makes them a challenge to process. This results in modification of processing equipment [2].

Most biodegradable polymers can be processed using a variety of processing methods. Following are the list of processing methods that will be discussed:

1. Injection molding
2. Compression molding
3. Film casting
4. Blow molding
5. Blown film extrusion
6. Thermoforming
7. Fiber spinning

Introduction to Bioplastics Engineering.
DOI: http://dx.doi.org/10.1016/B978-0-323-39396-6.00007-5

7.2 Injection Molding

Injection molding is the most widely used process for thermoplastic articles, especially for those that are complex in shape and require high dimensional precision. Biodegradable materials process like traditional thermoplastics such as PC or ABS and can be run on conventional machines as shown in Fig. 7.1 [3]. A typical general-purpose screw is shown in Fig. 7.2 [4]. High-shear screws such as a nylon screw are not recommended as they can generate a lot of shear heating. Much like standard thermoplastic, the recommended shot volume should be 30−80% of the barrel volume while hot spots are not recommended [5−7].

Although biodegradable polymers are semicrystalline and have low melting temperatures, they are relatively slow to crystallize. Cycle time and heat resistance can be improved by using nucleation technology. Biodegradable polymers tend to stick to metal surfaces during processing and absorb moisture on exposure to ambient air. This dilutes the effect of polymer drying. Biodegradable polymers are hygroscopic and must be dried to avoid a drop in molecular weight, melt viscosity, as well as increased potential for flashing and brittle parts. Since PLA and PHA are polyesters, drying requirements are stricter than for ABS, nylon, or Polycarbonate. Too high drying temperature would cause the material to get soft and agglomerate in the drying hopper. If it is too low, it will not dry as readily. The twin-bed desiccant dryers are not able to maintain the low temperatures as needed by biodegradable polymers, owing to temperature spikes after regeneration. Therefore it is recommended to use hopper agitation, fluid-bed crystallizers, or infrared crystallizing and drying units [7].

It is often recommended to purge the biodegradable polymer from injection molding machine after its use. The processing window of biodegradable polymers is very narrow. Therefore, it is tough to process biodegradable polymers between the melting point and the point of decomposition. Gels, black specs, and yellowing are created when the polymer is exposed to too much heat. These materials are hygroscopic and moisture sensitive thus, it is necessary that processors watch melt temperature, screw speed, injection speed, and proper drying. Polylactic acid (PLA) retains more heat than the standard polymer. Therefore it requires longer cooling time. It does not flow well in thin wall sections over long distances, and if additional pressure is added to fill, it increases shear, which can cause it to break down and become brittle [8,9].

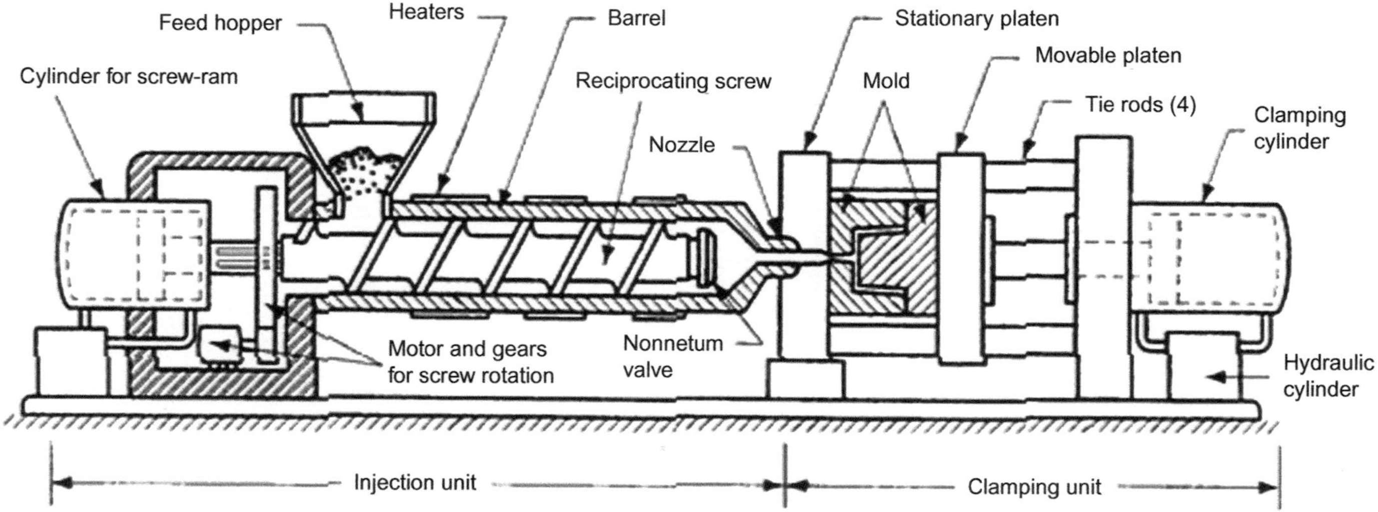

Figure 7.1 Conventional injection molding machine [3].

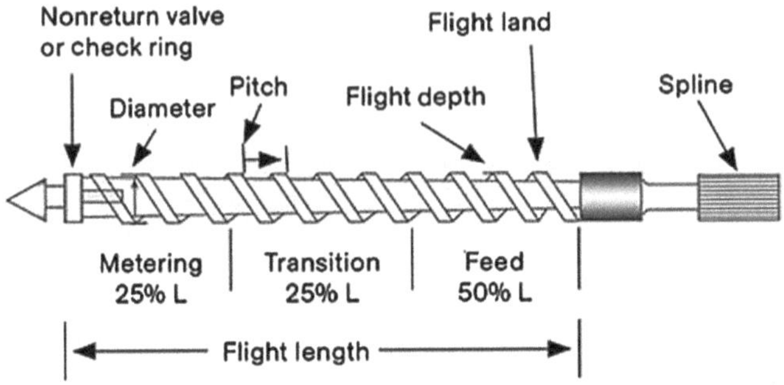

Figure 7.2 A typical general-purpose screw for injection molding [4].

PLA has a natural inclination to undergo lactide condensation on the cold tooling surfaces which can affect the surface finish and weight of the molded articles. This limits the minimum temperature that can be used during injection molding of PLA to 25–30°C. By polishing mold surfaces together with increased injection speed during fill can reduce the deposition of the lactide layer [10].

In this section, processing of some commercial biodegradable polymer will be discussed. These polymers are:

- NatureWorks Ingeo Series
- Novamont Mater-Bi
- DuPont Biomax TPS
- Metabolix Mirel
- TianAn Biologic
- Ecovio IS 1335

7.2.1 NatureWorks Ingeo Series

NatureWorks Ingeo 3000 series is designed for injection molding applications. It comprises of five resins with varying flow and transparency. Ingeo 3251D TDS is a higher flow and transparent material aimed at consumer electronics, cosmetics packaging, houseware, toys, and custom molding. It offers good optics, high clarity, high gloss, high modulus, good toughness, and UV transparency with no requirement of UV stabilizer. It can be molded with the same screws and molds used for PS, SAN, and ABS, though gating changes are required. Table 7.1 shows the comparison of different processing temperature profile of these resins. Grayed section indicates nonapplicable data [11].

Table 7.1 Processing Temperature Profile [11]

	3001D TDS	3052D TDS	3100HP TDS	3251D TDS	3260HP TDS
MFR, g/10 min (210°C, 2.16 kg)	22	14	24	80	65
MFR, g/10 min (190°C, 2.16 kg)				35	
Melt temperature (°C)	200	200	200	180–210	200
Feed throat (°C)	20	20	20	20	20
Feed temperature (crystalline pellets) (°C)	165	165	185	166–177	175
Feed temperature (amorphous pellets) (°C)	150				
Compression section (°C)	195	195	195	182–193	185
Metering section (°C)	205	205	200	188–205	190
Nozzle (°C)	205	205	200	188–205	190
Mold (°C)	25	25	120	25	120
Screw speed (rpm)	100–175	100–175	200	100–200	200
Back pressure (psi)	50–100	50–100	250	50–100	250
Mold shrinkage (in/in.)	0.004 ± 0.001	0.004 ± 0.001	0.017 ± 0.001	0.004 ± 0.001	0.017 ± 0.001

Table 7.2 Processing Conditions for Novamont Mater-Bi Injection Molding Biodegradable Polymer [12]

Desired moisture content (%)	0.4–2
Melt temperature (°C)	150–220
Nozzle temperature (°C)	194–204
Mold temperature (°C)	18–40
Shrinkage (in./in.)	0.003–0.01
Screw speed (rpm)	50–150

7.2.2 Novamont Mater-Bi

Novamont's injection grade of starch-based biodegradable polymer, Mater-Bi comes with MFI range from 6–30 g/10 min. It is recommended to process with a constant-taper, single-flighted screw having a 2.5:1 compression ratio and 25:1 L/D. A standard check ring can be used, along with medium to high injection speed. Table 7.2 shows processing conditions for Novamont's Mater-Bi biodegradable polymer. The melting temperature is in the range of 150–220°C, however, the maximum injection temperature must be less than 200°C and the residence time must be as short as possible. Molding of semicrystalline biopolymers can be up to 50% slower than more commonly used semicrystalline resins. Mold temperature of 18°C is typical [12].

7.2.3 DuPont Biomax TPS

DuPont Biomax TPS is a biodegradable polymer manufactured from chemically modified, high-amylose industrial cornstarch (Plantic technology) in five injection grades. These five grades are general-purpose and high-flow grades for thin wall parts, engineering, and two water-resistant grades requiring high stiffness. Two water-resistant grades can withstand water exposure for up to 4 or 12 weeks before biodegrading. Low-compression screws with a compression ratio of 2.2:1 to 2.8:1 and an L/D of 20:1 are recommended [12,13]. Processing conditions are shown in Table 7.3.

7.2.4 Metabolix Mirel

Mirel is a type of PHA (polyhydroxyalkanoate), primarily a blend of PHB (polyhydroxybutyrate) copolymers, additives, and mineral fillers. It is a type of semicrystalline thermoplastic polyester that is derived

Table 7.3 **Processing Conditions for DuPont Biomax TPS [12,13]**

Desired moisture content (%)	2–3
Melt temperature (°C)	165–190
Nozzle temperature (°C)	165–190
Mold temperature (°C)	25–50
Shrinkage (in./in.)	0.009–0.012
Screw speed (rpm)	50–150

Table 7.4 **Processing Properties of Mirel [14,15]**

	Mirel P1003/F1005	**Mirel P1004/F1006**
Processing recommendations		
Drying conditions (°C)	2–4 h at 80°C	2–4 h at 80°C
Melt temperature (°C)	160–165	160–165
Equipment recommendations		
Screw profile	2.2:1 to 2.6:1 (low shear GP)	2.2:1 to 2.6:1 (low shear GP)
Nonreturn valve	Standard check ring	Standard check ring
Processing conditions		
Barrel zone settings	**Reverse temperature profile**	**Reverse temperature profile**
Rear (°C)	175–180	175–180
Middle (°C)	170–175	170–175
Front (°C)	165–170	165–170
Nozzle (°C)	165–170	165–170
Mold temperature (°C)	55–65	55–65
Screw speed, slow (rpm)	<200	<200
Back pressure, low (psi)	<500	<500
Second stage pressure, low (psi)	<30% of first stage pressure	<30% of first stage pressure

from microbial fermentation of corn sugar. Mirel is supplied by Telles, a joint venture of Metabolix, Inc. and Archer Daniels Midland Co. Mirel has four injection molding grades. P1003 and P1004 are general-purpose injection molding materials with high modulus and high toughness respectively, and F1005 and F1006 are FDA cleared materials for use in nonalcoholic food contact applications. Table 7.4 shows differences in their processing properties [14,15].

All Mirel biodegradable polymers are thermally sensitive and must be dried to low moisture content (0.1%). They are processed using a reverse barrel-temperature profile ranging from 175–180°C in the rear to 165–170°C at the nozzle. Screw speed is recommended in the range of 50 to 150 rpm. Mirel has a recommended melt temperature of 160°C, and it decomposes above 180°C so control of melt temperature is vital [12].

7.2.5 TianAn Biologic

TianAn Biologic's PHBV (polyhydroxybutyrate-valerate) is a bio-polyester produced via bacterial fermentation of plant starches. TianAn offers Enmat Y1000P injection grade powder and pellets and Enmat Y5010P pelletized blend of PHBV and BASF's Ecoflex biodegradable (but not bio-based) resin. Both materials are approved for food contact in Europe and are approved by the Biodegradable Products Institute (BPI), N.Y.C., for composting [12]. Table 7.5 shows processing conditions for TianAn Biologic's Enmat products.

7.2.6 Ecovio IS 1335

The recommended processing temperature for different grades of Ecovio IS ranges from 180°C to 205°C. The choice of this range is dependent on the length of flow path and wall thickness as well as residence time. The optimum machine setting as determined by experience should start with a temperature of 195°C. A level or slightly rising temperature profile is advisable as shown in Fig. 7.3. Extremely high material temperatures and too long residence times are not desirable and can lead to decomposition of the mold material [16].

Table 7.5 Processing Conditions for TianAn Biologic's Enmat Products [12]

	Y1000P & Y5010P
Feed temperature (°C)	135
Compression section (°C)	145
Metering section (°C)	155
Adapter (°C)	160
Die (°C)	160
Screw speed (rpm)	40–50
Mold temperature (°C)	60

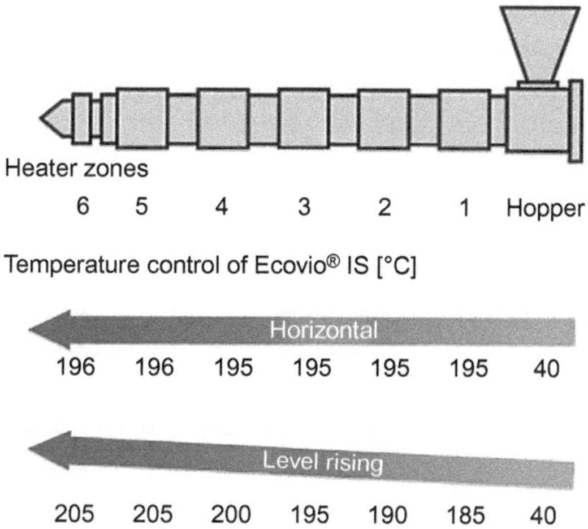

Figure 7.3 Examples of temperature control in the barrel [6].

Ecovio IS 1335 uses a single-flight, normal-cut, three-zone screw for its processing. However, shallow-cut screws are sometimes advantageous and provide even temperature distribution in the melt. For the processing of Ecovio IS grades, only wear-resistant steels should be used for cylinder, screw, and nonreturn valve [6].

Hot runner systems and heated nozzles are recommended for Ecovio IS. The externally heated systems offer greater operating safety because of the homogeneous melt temperature and a reliable purging effect. It is important to have good thermal isolation at the gate and by doing this, the heated and cooled sections can be controlled and the total energy demand for heating and cooling is reduced. Cold distributors are likewise possible, but not the preferred choice because of high temperature loss over the flow path. The type of gate that is most suitable depends on the particular application and must therefore be tested in each individual case. Temperatures between 10°C and 40°C should be selected for the mold temperature. Processing with a mold temperature of 25°C is to be preferred [12].

7.3 Compression Molding

Composites can be made by compression molding of PLA reinforced with *kenaf fiber* as well as *cotton fibers*. PLA pellets can be converted into PLA films by compression molding at a temperature of 180°C. Compression molding is a process in which two matched metal molds are used to fabricate composite product. Fig. 7.4 shows a compression molder in which the base plate is stationary while the upper plate is movable. Reinforcement and matrix are placed in the metallic mold and the whole assembly is kept in between the compression molder. Reinforcement in this case could be fiber or support material while matrix is biodegradable polymer. The two halves of the mold are brought together and heat and pressure are applied. The polymer placed in between the molding plates flows due to application of pressure and heat and acquires the shape of the mold cavity with high dimensional accuracy. Curing of the composite may be carried out either at room temperature or at some elevated temperature. After curing, the mold is opened and composite product is removed for further processing [17].

The controlling parameters in compression molding method to develop superior and desired properties of the composite are shown in Fig. 7.5 [17]. They are time, temperature, and pressure and all of them

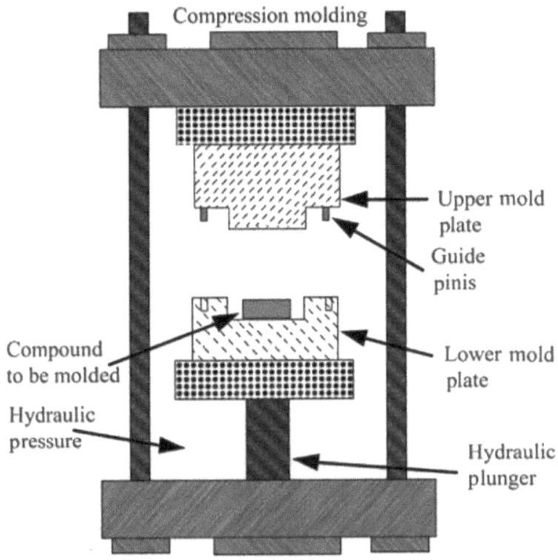

Figure 7.4 Compression molding arrangement [17].

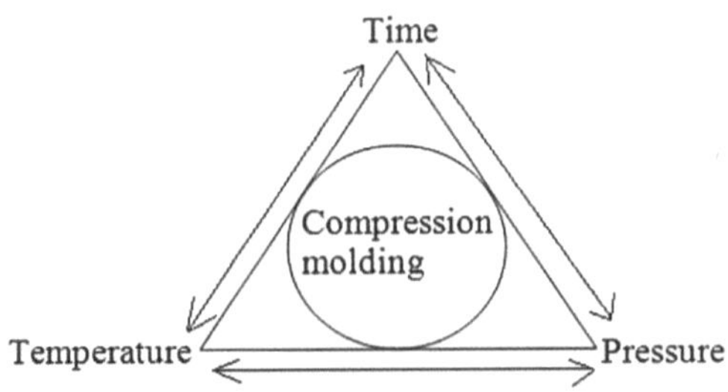

Figure 7.5 Process parameters of compression molding [17].

are critical and have to be optimized. If time is not sufficient, whether it is too high or too low, it can lead to defects. If temperature is too high, properties of fiber may get changed. If temperature is low than desired, fibers may not get properly wetted due to high viscosity of polymers especially for thermoplastics. If applied pressure is not sufficient, it will lead to poor interfacial adhesion of fiber and matrix. If pressure is too high, it may cause fiber breakage, expulsion of enough resin from the composite system. The other manufacturing factors such as mold wall heating, closing rate of two matched plates of the plates, and de-molding time also affect the production process.

Biodegradable polymers such as PLA, polyvinyl alcohol (PVA), soy-based plastics, and starch-based plastics can be effectively compression molded into products. Fig. 7.6 shows products made using compression molding [18,19]. Fig. 7.6a shows a tractor hood made by compression molding. Tecnaro GmbH successfully manufactured speakers shown in Fig. 7.6b by compression molding utilizing Arboblend resins. Arboblend combines lignin, biopolyesters, cellulose, biopolyamide, and bioolefins.

7.4 Film Casting

Biodegradable polymers have been successfully extruded using conventional extruders. A typical film casting process setup to process biodegradable polymer is shown in Fig. 7.7 [20].

Fig. 7.7 shows a typical film casting process. The casting equipment consists of five basic units:

- Extruder
- Forming unit—adapter and feed block
- Casting unit
- Winder unit
- Automation unit

The molten polymer exiting the die is quenched on a series of highly polished chrome rollers that are cooled with circulating water. Following the subsequent chill rolls, the film thickness distribution is determined by a thickness measuring unit. The film can be corona-treated if needed using electrical discharge process increasing surface tension and thus facilitating printing [20,21].

External deckles are used on the die to adjust the film width because of the thermal sensitivity of biodegradable polymers. However, this

(a) (b)

Figure 7.6 Compression molding of (a) tractor hood and (b) speakers [18,19].

should be avoided since the degraded resin behind the deckles can lead to edge instability. Horizontal roll stack configuration is recommended because of the low melt strength of biodegradable polymers. In the case of PLA, roller temperatures are set relatively higher to 25–50°C to prevent condensation of lactide monomers and slippage of web on the rollers. The die is positioned as close as possible to the entrance nip

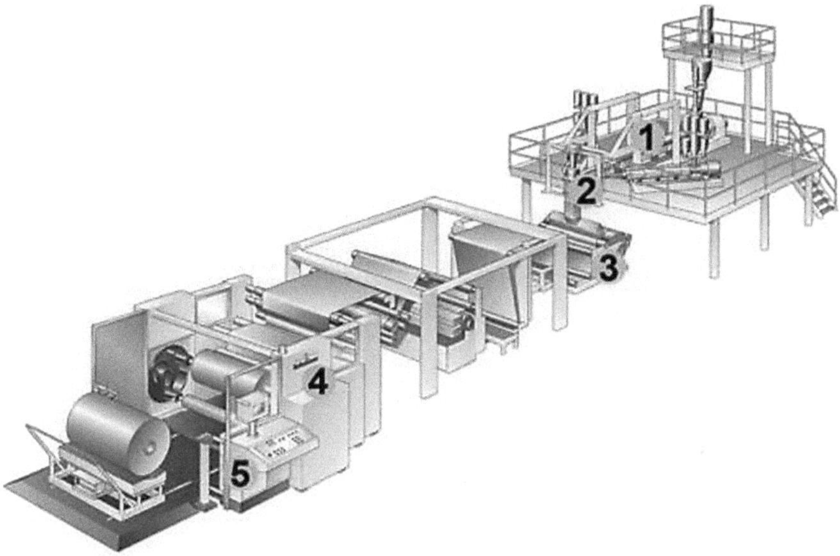

Figure 7.7 A typical film casting process—(1) extruder, (2) forming section—adapter and feed block, (3) casting section, (4) winder, and (5) automation [20].

Figure 7.8 Molten PLA extrudate coming out from the die [21].

however keeping it slightly higher than the nip can help in accommodating drooping of the molten polymer avoiding any chance of trapping air and reducing film defects [20]. As an example, Fig. 7.8 shows molten PLA extrudate coming out from the die [21].

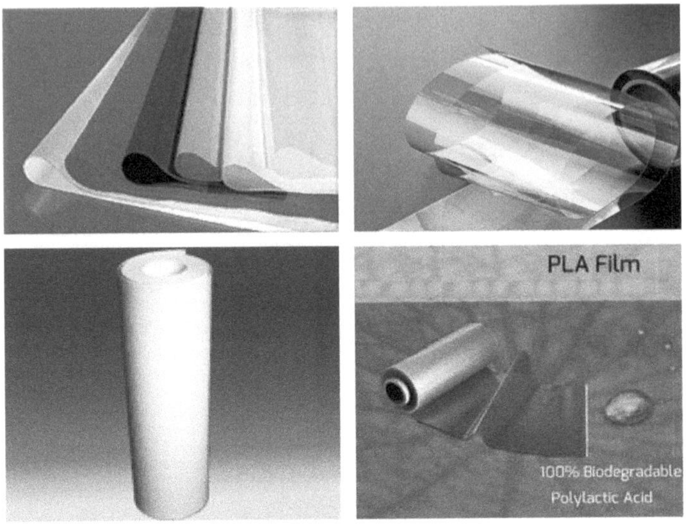

Figure 7.9 Cast film product examples [22–25].

The film is wounded on to a series of rolls using a contact/surface winder with an option for automatic roll change. The casting process can be automated making sure that the meter weight and thickness distribution remain constant. Fig. 7.9 shows examples of different types of films made from PLA using film casting process [22–25].

7.5 Blow Molding

Blow molding is the most common process best suited for hollow parts such as plastic bottles with uniform wall thicknesses, where the outside shape is a major consideration. There are two main types of blow molding [1]:

1. Injection blow molding
2. Extrusion blow molding

Injection blow molding is used for the production of hollow objects in large quantities. The main applications are bottles, jars, and other containers. A typical injection blow molding process is shown in Fig. 7.10. The process consists of four stages [26]:

- Preinjection stage — polymer is fed into the extruder and plasticizes plastic pellets into molten polymer.

- Injection stage — the molten polymer is fed into a manifold where it is injected through nozzles into a hollow, heated preform mold. The preform mold forms the external shape and is clamped around a core rod which forms the internal shape of the preform.
- Blowing stage — the preform mold opens and the core rod is rotated and clamped into the hollow, chilled blow mold. The core rod opens and allows compressed air into the preform, which inflates into the finished product shape.
- Ejection — after the cooling cycle, the blow mold opens and the core rod is rotated to the ejection position. The finished product is ejected off the core rod.

Extrusion blow molding process is slightly different than injection blow molding process. A typical extrusion blow molding process is shown in Fig. 7.11. In this process, the polymer is plasticized

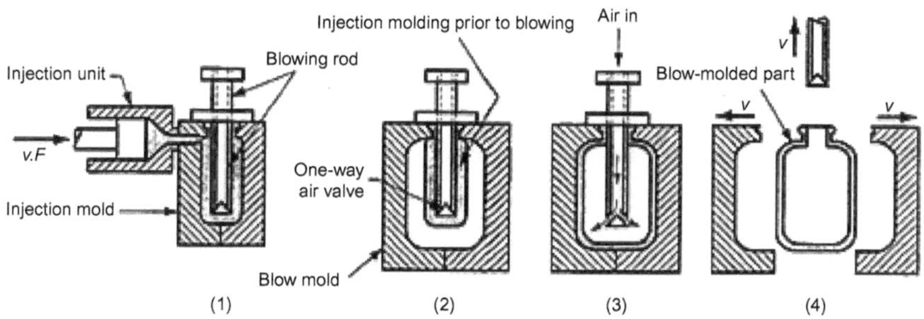

Figure 7.10 Injection blow molding process [26].

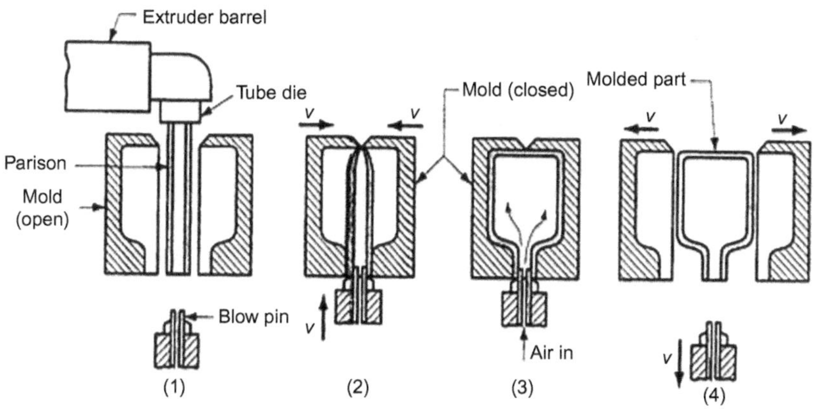

Figure 7.11 Extrusion blow molding process [26].

(a)

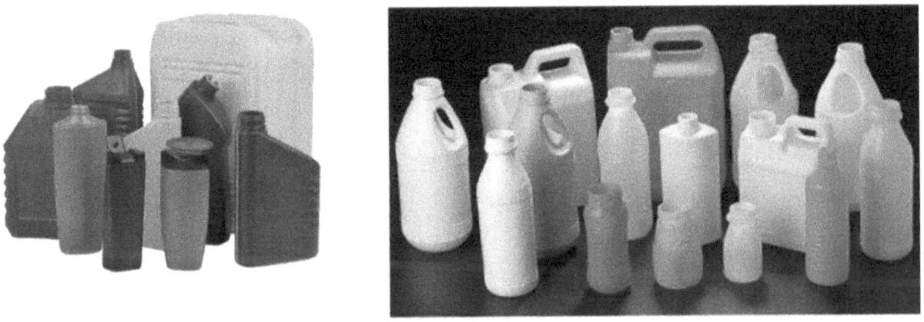

(b)

Figure 7.12 Examples of blow molding products made from (a) injection molding process and (b) extrusion blow molding process [27–30].

in a standard extruder and screw assembly. The molten polymer which emerges out through a die as a hollow section is known as parison. When a parison reaches a predetermined length, a hollow mold is closed around it. The mold mates closely at its bottom edge thus forming a seal. The parison is cut at the top by a knife prior to the mold being moved sideways to a second position where air is blown into the parison to inflate it or to the shape of the mold. After cooling cycle, the mold is opened and the product is ejected [26].

Fig. 7.12a and b shows some examples of injection blow molding and extrusion blow molding products made from biodegradable polymers [27–30]. FKUR has a list of biodegradable polymers that are used in blow molding applications [31]. FKUR's bio-flex grade material is used for blow molding applications. The BIO-FLEX is a blend of copolyester and PLA with, depending on the particular grade, a very high content of natural resource material. BIO-FLEX does not contain any starch or starch derivatives [31].

7.5.1 Stretch Blow Molding

Stretch blow molding is a modification of injection blow molding process and is also referred to as injection stretch blow molding

(ISBM). Stretch blow molding is divided into two different categories single-stage and two-stages. Single-stage process is done using one machine meanwhile in a two-stage process; preforms are premade prior to the blow molding process [31].

7.5.1.1 Single-Stage Stretch Blow Molding

- Parison is injected from the extruder into preform mold where the plastic is cooled to form preform.
- The preform is then reheated and placed in the bottle mold.
- The preform is stretched by extending the core rod.
- Compressed air is then blown into the stretched parison to expand to the shape of the mold.
- Part is cooled and ejected.

7.5.1.2 Two-Stage Stretch Blow Molding

- Premade preforms are placed in preform heaters to allow then to get soften.
- The mold is opened allowing softened preform to be inserted.
- Closing of the mold.
- Stretching of the softened preform.
- Compressed air is then blown into the stretched preform to expand to the shape of the mold.
- Part is cooled and ejected.

Fig. 7.13 shows a two-stage stretch blow molding process [32].

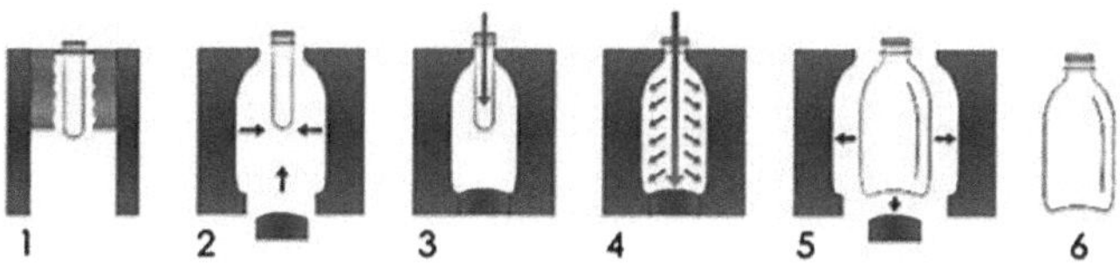

Figure 7.13 Two-stage stretch blow molding process—(1) preform heating, (2) opening of blow mold and insertion of the preform, (3) closing of the blow mold, (4) stretching and blowing, (5) cooling and part ejection, and (6) final part [32].

FKUR's Bio-Flex F6510 grade is used in stretch blow molding process because of its rigidity. Bio-Flex F6510 is predominantly composed of natural resource raw materials [33]. NatureWorks Ingeo 7001D biopolymer is used for ISBM bottle applications. Husky's first compostable water bottle made from NatureWorks PLA corn-based resin to the market uses ISBM process shown in Fig. 7.14 [34].

Figure 7.14 Husky's first compostable bottle made from injection stretch blow molding (ISBM) process [34].

7.6 Blown Film Extrusion

The blown film extrusion process is now transitioning from manufacturing films made from standard polymer to biodegradable polymers. A typical setup is shown in Fig. 7.15 [31]. In this process,

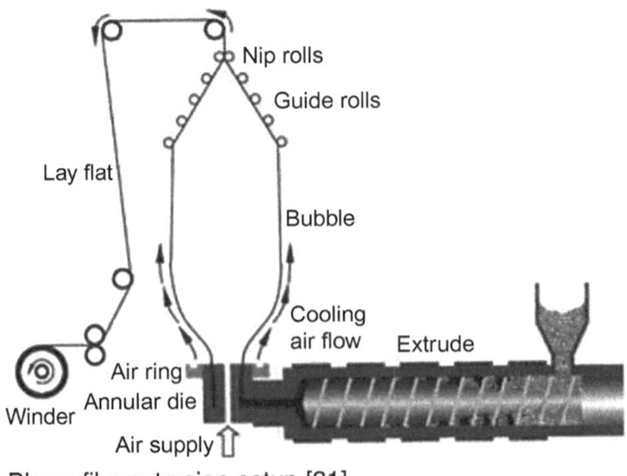

Figure 7.15 Blown film extrusion setup [31].

the polymer extrudate exits from the extruder head into the die. The molten polymer flows around the mandrel and emerges through a round opening (also called a ring) to form a tube. Air is then blown in through the die head into the ring, thereby inflating the tube into a thin tubular bubble. This tube is then flattened at the nip rolls and then taken to the winder. The blow-up-ratio (BUR) is defined by:

$$\text{BUR} = \frac{\text{Bubble Diameter}}{\text{Die Diameter}}$$

BUR ratios of 2:1–4:1 with the die temperature of 190–200°C are used for blown film extrusion of PLA films. By varying the BUR, screw speed, air pressure, and winder speed, films of different thicknesses (10–150 μm) and a degree of orientation can be achieved [31].

A biodegradable polymer such as PLA has weaker melt strength, and therefore, the formation of a stable bubble during extrusion is more difficult. Depending on the melt temperature, it could also limit the overall rate of gauge film that can be produced. Fig. 7.16 shows different applications of blown film extrusion [35–37].

7.7 Thermoforming

A typical thermoforming process begins when a plastic sheet is heated slightly above the glass transition temperature, for amorphous polymers, or slightly below the melting point, for semicrystalline materials. Although both amorphous and semicrystalline polymers are used for thermoforming, the process is the easiest with amorphous polymers because they have a wide rubbery temperature range above the glass transition temperature [38].

The heating is achieved using radiant heaters until the temperature reaches the forming temperature of the sheet. Once the sheet has been heated, it is forced against the mold-cavity contours, either pneumatically or mechanically. One way is by applying a vacuum in the mold cavity, which stretches the sheet until it touches the mold surface. The main issue with this forming process is the irregular thickness distribution across the part and is often corrected by using a mechanical plug [38].

Biopolymers have been successfully thermoformed into disposable cups, single-use food trays, lids, clamshells, and blister packaging as shown in Fig. 7.17 [39–43].

Fig. 7.18 shows the typical steps for thermoforming of PLA containers [44]. In this process, PLA sheet is heated to above glass transition

Figure 7.16 Blown film extrusion applications [35–37].

temperature (T_g) to make the sheet formable, but low enough to prevent the sheet from sagging. The heated sheet is then forced either pneumatically and/or mechanically against the mold, and then trimmed to form the final parts. PLA sheet can be thermoformed with vacuum, compressed air/vacuum, or only compressed air assistance. Depending on the part design, plug assist stretching or prestretching may be used to achieve more uniform part thickness [44].

In a thermoforming process, PLA sheet is typically heated using infrared (IR) radiation from the heating elements however, conduction and convective heating can also be used. Depending on the polymer's optimum IR absorbance frequency, the IR heaters should be set at

Figure 7.17 Examples of thermoformed biopolymer products [39–43].

temperatures where maximum energy is absorbed by the polymer. Set temperatures when thermoforming biodegradable polymers are significantly in the range of 80–110°C. Because of its heat transfer properties, aluminum molds are used to thermoform biodegradable polymers. Mold, trim tools, and ovens designed for thermoforming of PET, high impact polystyrene (HIPS), and oriented polystyrene (OPS) can be used for forming biodegradable containers [44].

7.8 Fiber Spinning

The process of producing fibers is called spinning. The two main types of spinning processes are:

- Melt spinning
- Electrospinning

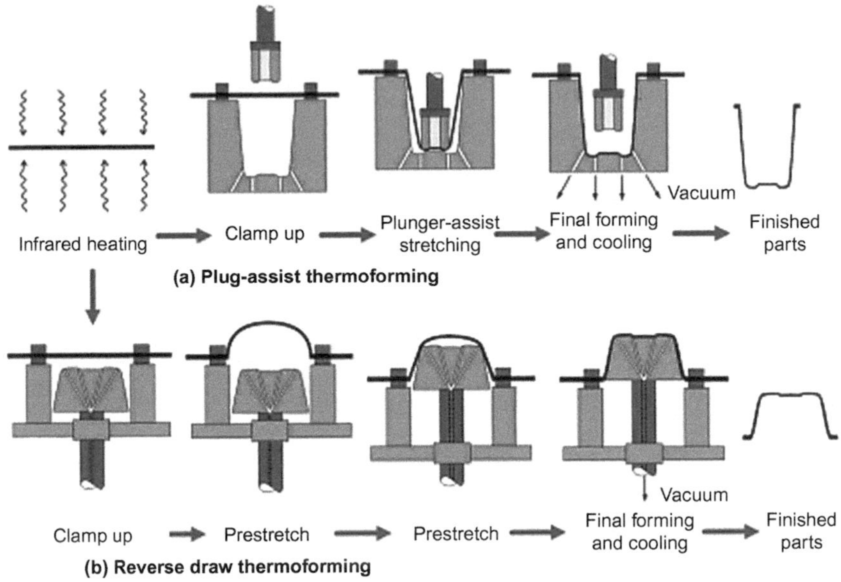

Figure 7.18 Thermoforming of heated PLA sheet over a female mold using plug-assist drawing and a male mold using reverse drawing processes [44].

Melt spinning is used for polymers that can be melted easily. Solvent-based spinning or electrospinning involves dissolving the polymer into a solution that can either be evaporated or when the solvent cannot be evaporated and must be removed by chemical means.

7.8.1 Melt Spinning

Fig. 7.19 shows a typical melt spinning setup consisting of extruder, a melt distribution unit (manifold), a metering pump which helps to regulate polymer flow rate, a spin pack to filter and extrude the polymer through fine holes, a quench chamber to cool and solidify the extruded polymer filament, and a winder. The entire line from extruder output to the spin pack is maintained at a constant temperature, which is called the spinning temperature [44].

A general-purpose single screw extruder 24 to 36:1 L/D with feed throat cooling is acceptable to process biodegradable polymers. Typically, a mixing tip is recommended along with static mixers in the product line to ensure temperature uniformity as well as optimum additive dispersion and melt polymer homogeneity.

The extruder melts the polymer and then compresses the fluid to remove any trapped gasses including air/nitrogen that is drawn along

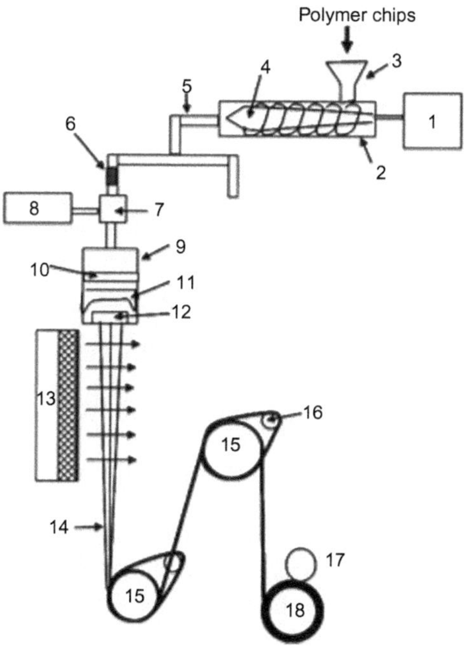

Figure 7.19 Schematic representation of melt spinning setup: (1) extruder drive, (2) extruder, (3) hopper, (4) screw, (5) manifold, (6) static mixer, (7) metering pump, (8) metering pump drive, (9) spin pack, (10) mesh filters, (11) distributor, (12) spinneret, (13) cross-flow quench chamber, (14) freshly spun yarn, (15) godet, (16) idler roller, (17) friction-driven winder, (18) yarn bobbin [44].

with the fed polymer chips. It also acts as a polymer fluid pump and provides the necessary pressure that is required for the polymer to flow from the extruder to the metering pump. The metering pump regulates the throughput of the polymer from the spinneret. The throughput rate and the winding speed together determine the denier of the spun filament.

Fig. 7.20 shows cross-sectional view of the spin pack which is the heart of the spinning system [44]. The spin pack contains a reservoir of polymeric fluid that provides a dampening effect on the gear metering pump. It also has a filter pack that removes the solid particles from the melt. The solid particles may include polymer gels, agglomerated additives, and contaminants. Once the polymer is filtered, it reaches the distributor. The distributor ensures proper supply of the polymer melt to all the spinneret holes [44].

Spinneret plate, a thick metal plate with fine channels drilled through it with an angle at the entry, allows the entry of the polymer melt with less pressure drop. This angle at the entrance facilitates polymer flow

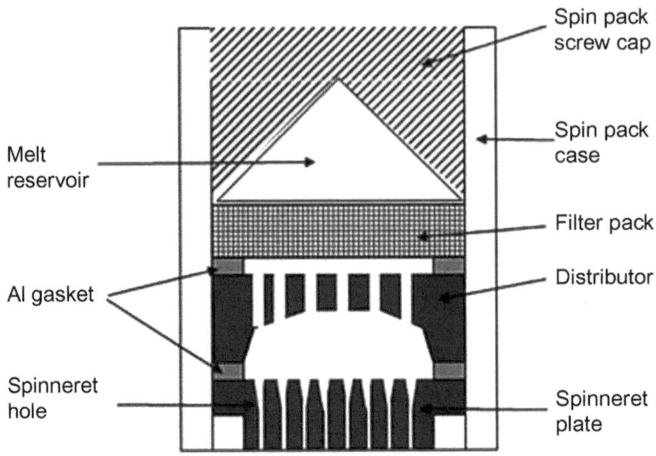

Figure 7.20 Cross-sectional view of the spin pack [44].

with reduced resistance. The extruded filament is stretched as it passes through an air quench zone. In this zone, fibers get solidified on making contact with the cool air. The fiber yarn is first passed through a take-up godet or a set of godet rollers before getting wound on the winder. This breaks the vertical path of the spinning and allows the winder to be adjusted easily in the available space [44].

Austria-based Lenzing AG offers Viscose and Lyocell, regenerated cellulosic fibers for various nonwoven applications such as electrical insulation, automotive and filtration, as well as medical drapes, swabs, and wound dressings. NatureWorks Ingeo PLA offers a renewable and biodegradable alternative to petrochemical-based fibers. Some of the applications include spunlaced wipes and hygiene products, in which PLA is blended with cotton or viscose, agricultural textiles, needle punched carpet and automotive products, spunbond filtration and geotextile products. Finland-based Ahlstrom Corp. has designed the unique tea bag that uses spunbond Ingeo PLA fiber as shown in Fig. 7.21 [45].

7.8.2 Electrospinning

Electrospinning is an approach to produce fibers from a polymer solution under the influence of electrostatic forces. Typical electrospinning equipment is made up of four main components as shown in Fig. 7.22 [44,46]:

1. DC power supply capable of supplying greater than 10 kV
2. Spinneret where the fiber jet is formed

Figure 7.21 The unique tea bag that uses spunbond Ingeo polylactic acid (PLA) fiber [45].

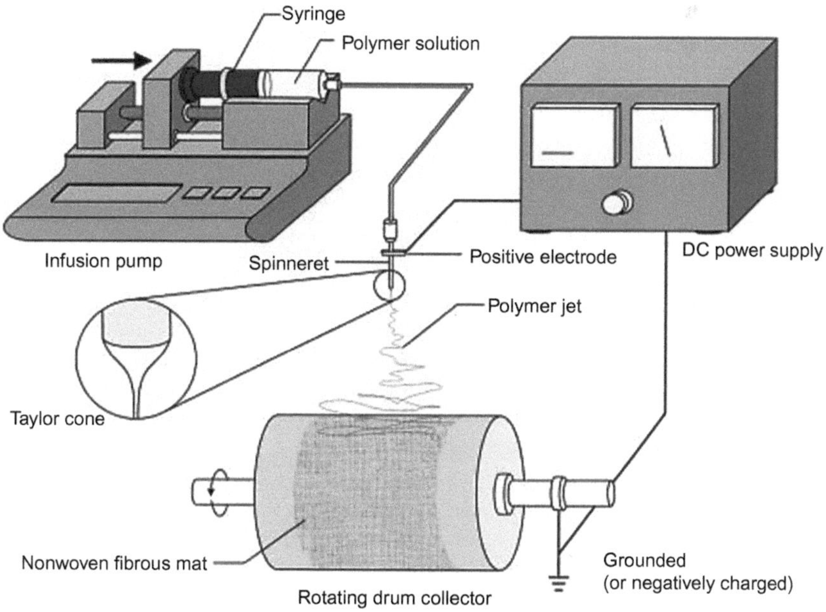

Figure 7.22 A typical electrospinning setup [44,46].

3. Syringe pump to deliver polymer solution to the spinneret
4. Collector, also acting as a counter electrode

When a polymer solution is charged, the resulting electrostatic force causes the droplet at the tip of the spinneret to elongate and form a Taylor cone. When the voltage reaches a certain critical level, the electrostatic forces overcome the surface tension of the solution, resulting in a continuous jet that ejects toward the collector. As the polymer jet takes flight in the air, the solvent vaporizes rapidly, producing a fibrous nonwoven laid on the collector. In the electrospinning process, the bending instability arising from the repulsive charge on the jet surface continuously stretches the charged jet, generating fibers with diameters as small as tens of nanometers. The bending instability, however, also causes the deposited fibers to be collected in a random pattern on the target, forming a nonwoven fiber mat [46].

Spinnability, fiber diameters, fiber uniformity, defects, and fiber alignment are affected by polymer properties as well as electrospinning process setup. The disordered structure greatly limits their application in areas requiring well-aligned and highly ordered architectures, such as device fabrication in nanoelectronics and photonics, reinforcements, and electrochemical sensors. Thus, one of the biggest challenges in this area is how to obtain highly ordered and controllable fiber mats and how to assemble them onto the device system [46–51]. Table 7.6 lists parameters that affect how well fibers can be produced.

A wide range of biodegradable polymers have been successfully electrospun into nanofibers [52], including natural proteins such as collagen, gelatin, silk, chitosan and alginate; synthetic polymers such as polyglycolide (PGA), poly(ε-caprolactone) (PCL), PLA and their copolymers P(LLA-CL), and PLGA that have been approved by FDA for

Table 7.6 Electrospinning Parameters

Polymer properties	Electrospinning setup properties
1. Polymer concentration 2. Viscosity of the solution 3. Molecular weight of the polymer 4. Surface tension of the solution	1. Flow rate 2. Type of the collector 3. Needle size 4. Electric field on the syringe and the collector

clinical use; and their blends [52,53]. Fig. 7.23 displays the biological, mechanical, and physiochemical properties of commonly studied biodegradable natural and synthetic polymers [53].

Electrospun fibers made from biodegradable polymers are primarily used for tissue engineering applications. They are used in blood vessels and nerve regeneration, scaffold or tissue regeneration, and porous membrane for skin regeneration as shown in Fig. 7.24 [54–56].

In addition, electrospun fibers are also used in other applications such as:

- Filtration — Liquid, gas, and molecule filtration
- Energy and Electronics — LCD devices
- Sensors — Thermal, piezoelectric, fluorescence optical sensor
- Protective clothing — Antibiochemical gases, trapping aerosols
- Medical — Drug delivery, wound dressing, haemostatic devices

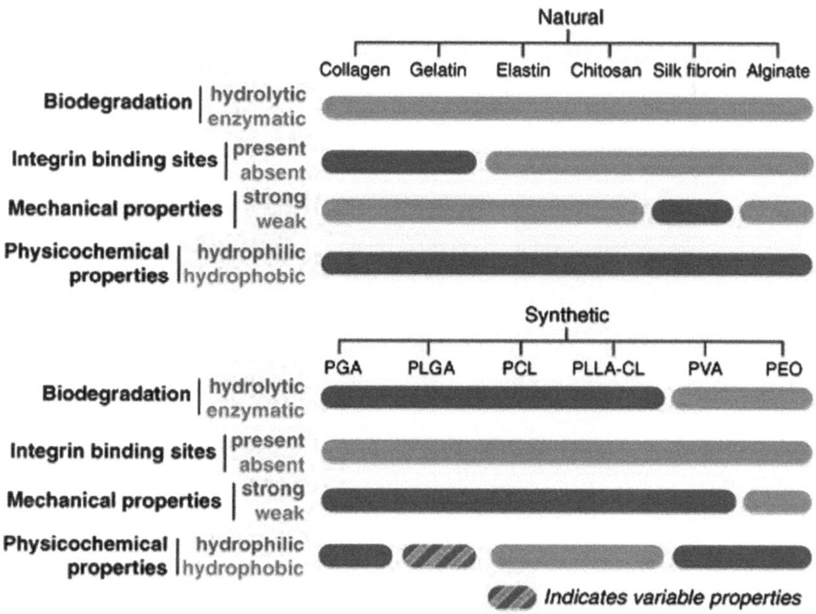

Figure 7.23 Biological, mechanical, and physiochemical properties of commonly studied biodegradable natural and synthetic polymer [53]. Reprinted with permission from Elsevier.

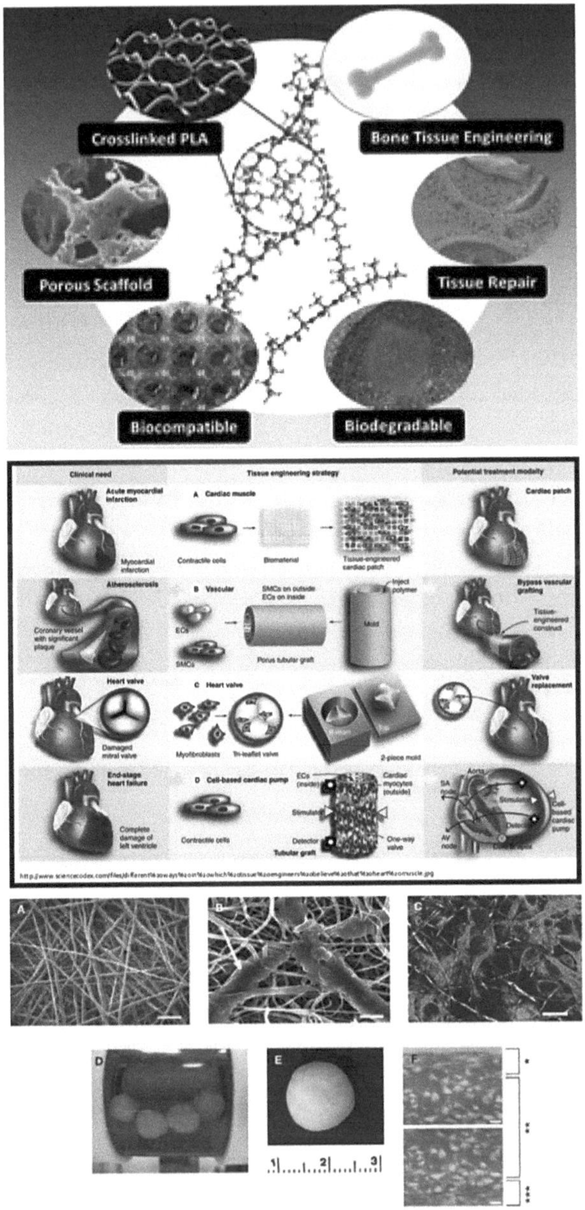

Figure 7.24 Electrospun fibers used in tissue engineering applications [54–56].

References

[1] D.K. Platt, Biodegradable Polymers: Market Report, iSmithers Rapra Limited, 2006.
[2] J. Xu, Microcellular Injection Molding, Wiley Publishers, 2011 (Chapter 4).
[3] Elite Machinery Systems LLC, <http://elitemachinerysystems.com/plastic_injection_molding.php>.
[4] M. Sepe, Working with color concentrates, Plastics Technology, April 2012.
[5] N. Navratilova, A. Naplava, Study of biodegradable plastics produced by injection molding.
[6] H. Rees, Understanding Injection Molding Technology, Hanser/Gardner Publications, Inc., Cincinnati, 1995.
[7] L.-T. Lim, R. Auras, M. Rubino, Processing technologies for poly(lactic acid), Prog. Polym. Sci. 33 (8) (2008) 820–852. Elsevier.
[8] V.R.N. Telis, Biopolymer Engineering in Food Processing, Table 8.4, CRC Press, 2012, 311.
[9] G.P. Felton, Biodegradable Polymers: Processing, Degradation and Applications, Nova Science Publishers.
[10] J.R. Dorgan, H. Lehermeier, M. Mang, Thermal and rheological properties of commercial-grade poly(lactic acid)s, J. Polym. Environ. 8 (2000) 1–9.
[11] NatureWorks 3 Series for Injection Molding Technical Resources, <http://www.natureworksllc.com/Technical-Resources/3-Series>.
[12] M. Knights, Injection Molding Biopolymers: How to Process Renewable Resins, Plastics Technology, 2009.
[13] S. Peters, Bioplastics Based on Thermoplastic Starch, Material Revolution: Sustainable and Multi-Purpose Materials for Design and Architecture, Walter de Gruyter, p. 37.
[14] Mirel™ P1003/F1005 Data Sheet Injection Molding Grade, <http://www.metabolix.com/sites/default/files/MirelP1003_F1005Datasheet042012.pdf>.
[15] Mirel™ P1004/F1006 Data Sheet Injection Molding Grade, <http://www.metabolix.com/sites/default/files/MirelP1004_F1006Datasheet032012.pdf>.
[16] Technical Datasheet, Ecovio® IS 1335, IDES.com, <http://catalog.ides.com/Datasheet.aspx?I=43838&FMT=PDF&E=276650>.
[17] Lecture 5.6: Compression Molding, Indian Institute of Technology Patna, <http://www.nptel.ac.in/courses/112107085/module5/lecture6/lecture6.pdf>.
[18] Ashley Industrial Molding, Custom Molded Hood Incorporating Part Consolidation for the Agricultural Industry, <http://www.ashinmold.com/custom-molded-hood-incorporating-part-consolidation-for-the-agricultural-industry.html>.
[19] J.H. Schut, New Technologies in Stretching and Touching, Plastics Engineering Blog, <http://plasticsengineeringblog.com/page/5/> 2013.
[20] B. Rieger, A. Kunkel, G.W. Coates, R. Reichardt, E. Dinjus, T.A. Zevaco, Synthetic Biodegradable Polymers, Springer, 2012.

[21] Bioplastics Magazine, <http://www.bioplasticsmagazine.com/en/online-archive/data/5714.php>.

[22] <http://www.directindustry.com/prod/ching-hsing-iron-works-co-ltd/product-75292-733061.html>.

[23] Medical Film Extrusion, Monolithic Industries, <http://medicalextrusion-technologies.com/film.html>.

[24] Biodegradation PLA/PBAT Cling Film Extrusion, <http://hbfilmmachine.com/product-2-5-pla-pbat-cling-film-line-en/134736>.

[25] Polylactic Acid (PLA) Biodegradable Films, <http://www.hellotrade.com/xanadu-technologies/product.html>.

[26] Blow molding, Sinotech, <http://www.sinotech.com/blowMolding.html>.

[27] Blow molded products, Valencia Plastics Inc., <http://valenciaplastics.com/>.

[28] Injection blow molding of sterilized polypropylene centrifuge bottles for the biopharmaceutical industry, BMP Biomedical Polymers, <http://www.biomedicalpolymers.com/injection-blow-molding-polypropylene-centrifuge-bottles-biopharmaceutical.html>.

[29] Fong Kee International Machinery, <http://www.fki.com/productDetail34b1.html>.

[30] Blow molding, DT Resources, <http://dtresource.com/blow-moulding.html>.

[31] J. Ren, Biodegradable poly(lactic acid): Synthesis, modification, processing and applications, Springer, 2010.

[32] Stretch blow molding, SMF Germany, <http://www.smfgmbh.com/en/offer/pet-blowers>.

[33] FKUR, Products, <http://www.fkur.com/produkte.html>.

[34] Husky helps develop world's first compostable water bottles, Husky Corporation, <https://www.husky.co/News.aspx?id = 163>.

[35] PLA Shrink label film for shrink sleeve labels, Allen Plastics Industries, <http://www.allenpack.com/en/2-1772-34353/product/PLA-Shrink-Label-Film-for-Shrink-Sleeve-Labels-id295949.html>.

[36] Biodegradable shrink wrap plastic films, <http://www.kempner.co.uk/shrink-wrap-plastic-film/biodegradable-shrink-plastic-films/>.

[37] Bio-Olefin, biodegradable shrink film, National Shrinkwrap, <http://www.nationalshrinkwrap.com/home.html>.

[38] S.A. Ashter, Thermoforming of Single and Multilayer Laminates: Plastics Films Technologies, Testing, and Applications, Elsevier, 2014.

[39] SeeShell PLA large deep hinged clamshell, BioMass Packaging, <http://biomasspackagingstore.com/seeshell-pla-large-deep-hinged-clamshell-9x9x3.aspx>.

[40] Taiwan Trade Company, <http://www.taiwantrade.com.tw/MAIN/en_front/searchserv.do?method=listProductProductDetail&catalog_id=777318&come_soon=0>.

[41] Bioplastics 9 lives, Sustainable manufacturer network, <http://sustainablemfr.com/materials/bioplastics-9-lives>.
[42] M. Knights, Processing Biopolymers for Rigid Sheet and Thermoforming, Plastics Technology, June 2011<http://www.ptonline.com/articles/processing-biopolymers-for-rigid-shee-thermoforming>.
[43] P. Reynolds, AT&T is sweet on sugarcane-based thermoform, Greener Package, <http://www.greenerpackage.com/renewable_resources/att_sweet_sugarcane-based_thermoform>.
[44] R. Auras, L.-T. Lim, S.E.M. Selke, H. Tsuji, Poly(Lactic Acid): Synthesis, Structures, Properties, Processing, and Application, Wiley, 2010.
[45] J.B. Rodie, Natural and renewable fibers, <http://www.textileworld.com/Issues/2010/September-October/Nonwovens-Technical_Textiles/Fibers_For_Nonwovens>.
[46] A.L. Yarin, S. Koombhongse, D.H. Reneker, Bending instability in electrospinning of nanofibers, J. Appl. Phys. 89 (5) (2001) 3018–3026.
[47] D. Kai, S.S. Liow, X.J. Loh, Biodegradable Polymers for Electrospinning: Towards Biomedical Applications, Materials Science and Engineering, C 45, Elsevier, 2014.
[48] D.H. Reneker, A.L. Yarin, F. Hao, S. Koombhongse, Bending instability of electrically charged liquid jets of polymer solutions in electrospinning, J. Appl. Phys. 87 (9) (2000) 4531–4547.
[49] D.H. Reneker, T. Han, Electrical bending and mechanical buckling instabilities in electrospinning jets. Responsive Soft Matter: Chemistry and Physics for Assemblages, Films and Forms, 948 (2006), pp. 29–40.
[50] A.L. Yarin, W. Kataphinan, D.H. Reneker, Branching in electrospinning of nanofibers, J. Appl. Phys. 98 (6) (2005). 64501-1–64501-12.
[51] H. Fong, I. Chun, D.H. Reneker, Polymer 40 (1999) 4585.
[52] N. Ashammakhi, A. Ndreu, A.M. Piras, L. Nikkola, T. Sindelar, H. Ylikauppila, et al., Biodegradable nanomats produced by electrospinning: expanding multifunctionality and potential for tissue engineering, J. Nanosci. Nanotechnol. 6 (2006) 2693–2711. American Scientific Publishers.
[53] S. Agarwal, J.H. Wenorff, A. Greiner, Use of electrospinning technique for biomedical applications, Polymer 49 (26) (2008) 5603–5621. Elsevier.
[54] Toyota Technological Institute, <http://www.toyota-ti.ac.jp/english/research/labolatories/mate/post-19.html>.
[55] R. Soto, Cardiovascular tissue engineering, BME501: Tissue Engineering, <http://www.slideshare.net/rsoto21/cardiovascular-tissue-engineering>.
[56] F.H. Chen, K.T. Rousche, R.S. Tuan, Technology insight: adult stem cells in cartilage regeneration and tissue engineering, Nat. Clin. Prac. Rheumatol. 2 (2006) 373–382.

8 Extrusion of Biopolymers

8.1 Introduction

Extrusion is a well-known process, developed to produce polymer products of uniform shape and density, and its industrial application dates back to the 1930s [1]. It is one of the most widely applied processing technologies in the plastic, rubber, and food industries and is used to prepare more than half of all plastic products including bags, films, sheets, tubes, fibers, foams, and pipes. In this process, under the application of heat and pressure, the polymer is melted and is forced though an orifice in a continuous process [1]. Biopolymers can be processed using conventional extrusion. However, the processing of biopolymers involves a complex interaction of various physical, chemical, and thermodynamic phenomena, making it very difficult to understand. The properties of the final product are influenced by a variety of factors like processing conditions and depend strongly on the inherent properties of the biopolymer being used.

Extrusion is carried out using an extruder containing one or two rotating screws that transport material down the barrel. Extruders consist of four distinct components:

1. A hopper and/or a secondary feeder fed with the material to be extruded and continuously supplied in a controlled manner entering into the barrel.
2. Conveying section which comprises the barrel and the screw(s) that transport, and where applicable, mix the material.
3. Adie which gives material the shape as it leaves the extruder.
4. Downstream auxiliary equipment for cooling, cutting, and/or collecting the finished product.

Based on the number of screws in the barrel, extruders can be classified as single and twin screw and are shown in Fig. 8.1 [2].

Single-screw extruders are primarily used for melting and conveying polymers to extrude them into continuous shapes. The screw is

Introduction to Bioplastics Engineering.
DOI: http://dx.doi.org/10.1016/B978-0-323-39396-6.00008-7

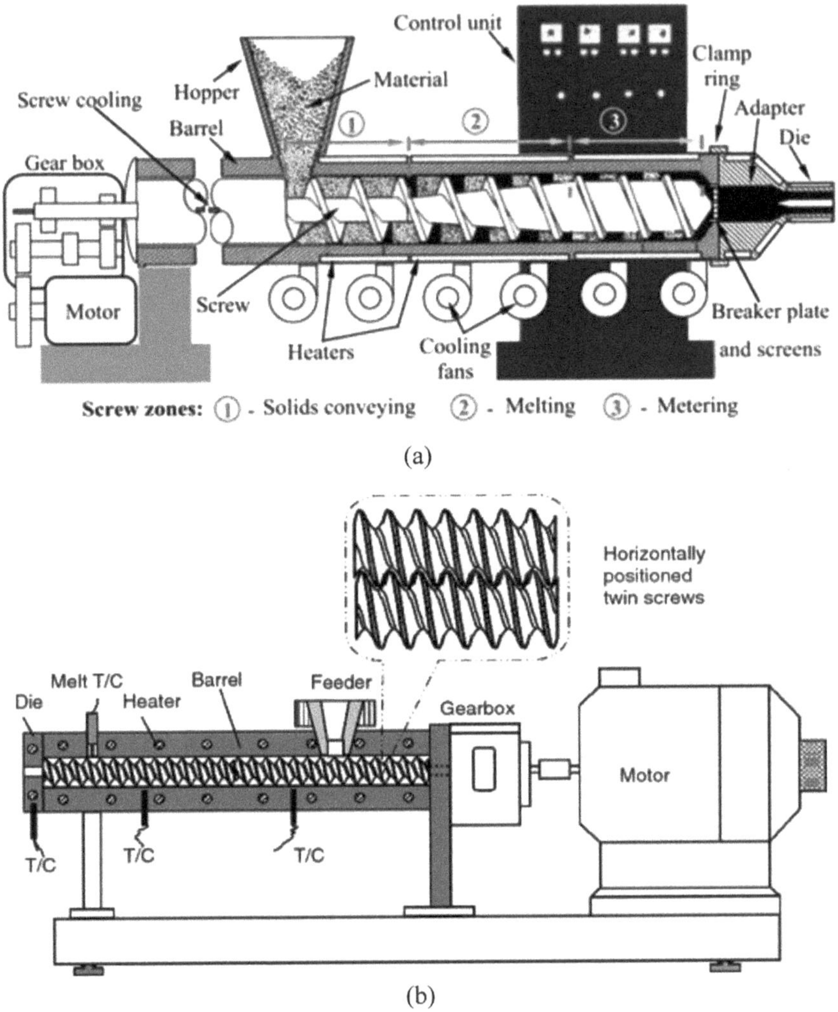

Figure 8.1 A schematic representation of (a) single-screw extruder and (b) twin-screw extruder (TSE) [2].

surrounded by the barrel that resists high pressures. The barrel usually has three or more heating zones with heaters mounted on the barrel to raise the barrel and screw to the required process temperature. When the process temperature is driven above the temperature set point, barrel cooling is triggered, either by air or by liquid which cools the barrel.

The polymer is melted by frictional heating within the barrel. In order to have low melt viscosity to allow conveying of melt down the barrel, proper mixing, and to avoid thermal degradation of the polymer,

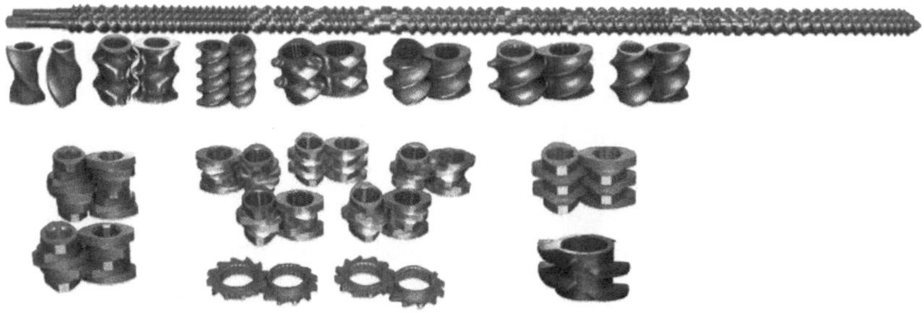

Figure 8.2 Mixing and conveying elements on extruder shafts [3].

the barrel temperatures are optimized. Once the polymer is melted and mixed, the melt flows through the breaker plate and a die, and thereafter is cooled into solid form [4].

The TSEs are used for melt compounding two different polymers or a polymer with additional additives such as stabilizers, lubricants, and plasticizers and for devolatilization. In this process, melting is accomplished when materials undergo shearing between the rotating screws and between the screws and the wall of the barrel as they are conveyed. The screws of a twin-screw extruder are usually to provide different types of mixing and conveying conditions at various zones in the barrel.

A complete modular screw with multiple elements fitted on a common shaft is shown in Fig. 8.2 [3]. This allows the tailoring and optimization of the screw design for each product. Sections of the screw can be designed to perform particle-size reduction, mixing, and conveying functions. The length of the screw in relation to the barrel diameter (the *L*/*D* ratio) is chosen to optimize the degree of mixing and the number of zones required to achieve the final product characteristics.

Processing polylactic acid (PLA) and other biopolymers is complicated because they are sensitive to heat and shear. The biopolymers are also prone to hydrolytic degradation if moisture is present. Other factors such as elevated temperatures, pressures, and moisture all resulted in a molecular weight loss and decreased mechanical properties. Torque is often the rate limiting factor when processing biopolymers. These polymers require a lot of torque to melt and process. All of these challenges must be addressed in the design and configuration of the twin-screw extrusion system [5].

8.2 Conventional Extrusion of Biopolymers

Historically, TSEs have been used to produce pellets to consistently feed into secondary processes such as an injection molding machine or a single-screw extruder. There is also a fast growing trend to bypass the pelletizing step and to produce a film, fiber, sheet, or profile from the TSE. This continuous is termed as direct extrusion. Either way, the operation performed by the TSE is identical.

Twin-screw extrusion is the preferred method for compounding PLA and other biopolymers. Compounding PLA on a TSE with masterbatches, impact modifiers, fillers, additives, and other polymers increases the functionality of PLA [5].

It is important to make sure that the extruder has been cleaned and purged before feeding PLA to avoid cross-contamination. Modular barrels and screws are used when processing using co-rotating, intermeshing TSEs. Modular barrel and screws allows specific screw and barrel geometries to be matched to the unit operation performed in the machine. Self-wiping feature of co-rotating TSEs allows the use of standard metallurgies for screws and barrels such as hardened tool steels and powder metallurgies for increased abrasion resistance. All other parts in the process where there are stagnant flow areas such as adapters, screen changers, and dies, should be made from stainless steel and hard-chrome plated for increased corrosion resistance.

For any TSE, free volume is an important design parameter. It is directly proportional to the ratio of outside diameter (OD)/inside diameter (ID) of the screws, defined by dividing the OD to the ID of each screw. Increased free volume is possible with a smaller screw shaft, but at the expense of attainable torque. Because torque is limited by the cross-sectional area of the screw shaft, the shaft design, and metallurgy, deeper screw flights result in less torque and more free volume, since a smaller diameter screw shaft is mandated. Shallower flight depths and lower OD/ID ratios result in a higher average shear rate, compared with lower shear rates inherent with deeper flights and a higher OD/ID ratio [5].

It was considered that the symmetrical splined screw shafts with a 1.55 OD/ID ratio provided the best balance between torque and volume. However, these shafts induce both tangential and radial force into the power transmission train. Thus, the resultant force is not optimized and the radial force is not applied in a beneficial direction. Recently, asymmetrical splined shafts were introduced that isolate the tangential force vector that results in higher torque transmission with a smaller shaft diameter. The asymmetrical shaft with 1.66 OD/ID ratio and deeper flight depths resulted in a higher free volume, and increased torque. These shafts are now considered as the optimum by many. The combination of deeper flights with increased torque is particularly beneficial for processing PLA compounds. Fig. 8.3 shows difference in two splined shafts [5].

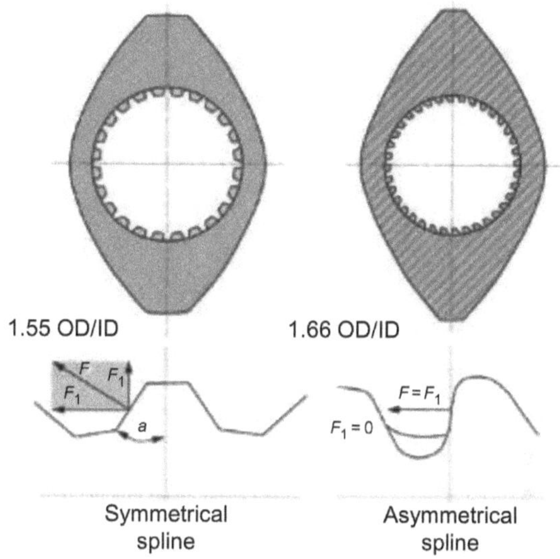

Figure 8.3 Asymmetrical splined shafts isolate the tangential force vector that results in higher torque transmission with a smaller diameter shaft. As a result, a 1.66 OD/ID ratio with deeper flight depths, higher free volume, and increased torque is now perceived as optimum when processing PLA [5].

The conventional extrusion has two distinct families of TSEs defined as:

1. Low-speed late fusion (LSLF) TSEs (run up to 50 rpm).
2. High-speed energy input (HSEI) TSEs (run up to 1200+ rpm).

HSEI TSEs are primarily used for compounding, reactive processing, and/or devolatilization. Because HSEI TSE are heat and shear insensitive, they are not used for biopolymer processing. In contrast, LSLF counterrotating TSEs are designed to mix at low shear and pump at uniform pressures. Heat- and shear-sensitive biopolymers such as PLAs are compounded using LSLF TSEs. These extruders are often inadequate to perform energy-intensive processing [5,6].

8.2.1 LSLF Extrusion

The LSLF counterrotating intermeshing mode is characterized by a gentle melting and mixing effect that ensures a narrow residence time in conjunction with a high-pressure buildup, as compared to the high-speed TSE that is specified for energy-intensive applications. This type

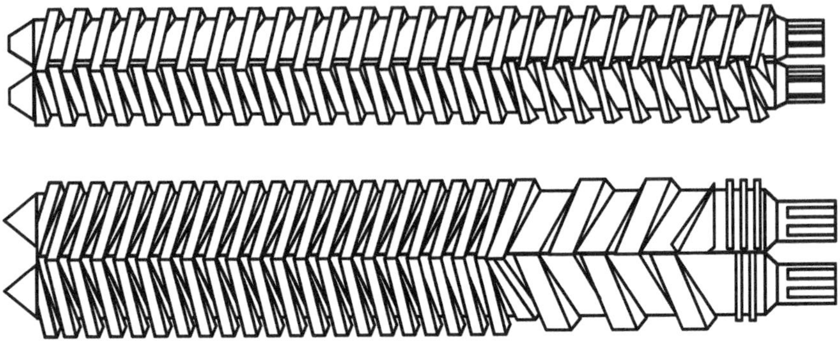

Figure 8.4 Low-speed, late fusion counterrotating screw sets [7].

of extrusion is used for applications where shear- or temperature-sensitive biopolymers are being processed, or where high head pressures are desired, and/or where the materials do not convey well by drag flow. Co-rotating designs can be made to operate similarly through the use of nontraditional screw geometries.

The screw flights coverage in the same direction in the nip region, causing the open spaces between the flights to be small, minimizing leakage from one screw channel to the next as shown in Fig. 8.4 [7].

8.3 Starch Extrusion

Starch can be processed either on a single-screw extruder or a TSE. A single-screw extruder can handle the high viscosity of starch and provide a high processing pressure for continuous metering of starch through die shapes. Individual barrel zone temperature control, multiple feeding, and screw configuration for a different degree of mixing gives TSE a large operational flexibility for intensive mixing and compounding of components into plasticized starch. One of the significant advantages of twin-screw extrusion over single-screw is that the twin-screws have a self-wiping ability minimizing problems with conveying the starch powder at the feeding port [8,9].

Typical starch extrusion equipment has eight temperature controlling zones along the barrels of the extruder as shown in Fig. 8.5 [10]. Extrusion of starch results in massive viscous heat dissipation which can increase the temperature of the plasticized starch by up to 50°C. Therefore, water cooling is necessary. Typically, the barrels have a longitudinal split design for sample collection and cleaning purpose, the extruder is capable to prevent the moisture loss during extrusion. By

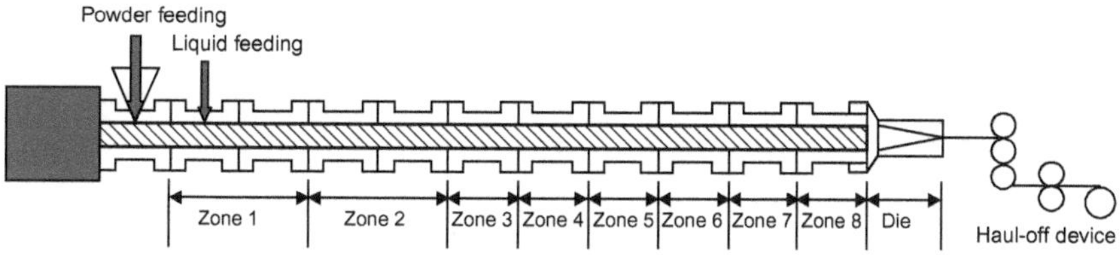

Figure 8.5 Schematic representation of a typical extrusion system to produce starch film [10].

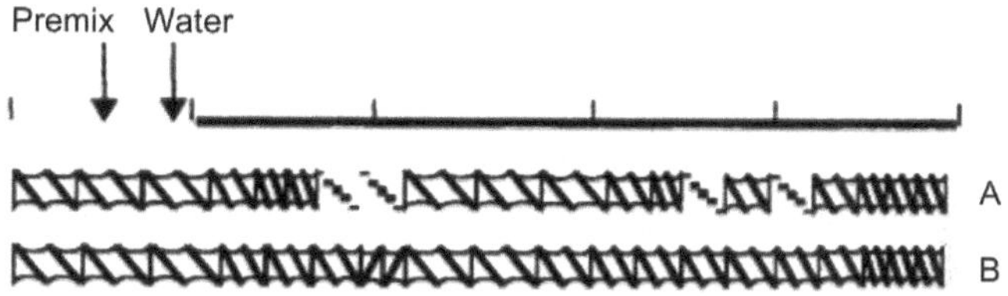

Figure 8.6 Screw designs used in starch extrusion (right-hand side is exit from barrel) [11].

coupling with different kinds of dies, the extruder can be used to produce variety of different products ranging from strands that are then blade cut into pellets or films/sheets which are then collected by a take-off rolls [10].

Fig. 8.6 shows the two screw configurations used for starch extrusion; with right-hand side is exit from barrel. Design A has different zones with positive kneader elements, whereas design B uses reverse screw elements. Using design B will result in an increase of the filled region and in a change of the pressure profile along the extruder axis [11].

The melting temperature of dry starch is often higher than its decomposition temperature, and therefore requires starch to be processed with water. Water acts like a plasticizer by lowering the melting temperature and plasticize the starch polymer. If water is the only plasticizer used in starch processing, it may result in unstable processing condition due to the water evaporation. Therefore, it is recommended to use lubricants such as magnesium stearate, calcium stearate, and fluoroelastomers to improve the processability of plasticized starch-based materials. The use of a lubricant during processing can reduce the tendency of the material to stick to the die and clog it [11].

8.4 Extrusion Screw Designs

The extruder screw is the heart of the machine. Everything revolves around the extruder screw, literally and figuratively. Extruder screws as a general rule employ simple-start flights although two or even more starts are sometimes used. The alteration of flight depth to obtain a compression on the material as it moves toward the die is standard practice on most screws and is either carried out progressively throughout the length of the screw, or in stages or by a combination of both methods. The deliberate reduction of the volumetric capacity of the screw channel or channels is necessary in order to accommodate the reduction in volume of the material as it becomes fluid and homogenous and to apply compression to the material so that the channel is completely filled [12].

If the screw is considered solely as a means of conveying the material, then the diameter and length are major factors in determining its volumetric capacity and hence the quantity of feed material can handle. These two functions remain important controls of capacity irrespective of screw design. The length and diameter have a second important influence in that they affect the rate at which heat is transferred from the barrel walls to the material, and this in turn affects the amount of heat generated by friction and shear, the energy input, and the power to throughput ratio. Different thermoplastic materials require different processing conditions, so that maximum screw efficiency for every material with one design of screw is not possible [13].

Each process task, such as solid feeding, is associated with a particular section of the extruder screw, although several tasks are sometimes accomplished at once.

There are three main types of screw designs popular among biopolymers:

1. Turbo-Screw
2. Solid-state shear pulverization
3. Reactive extrusion polymerization

8.4.1 Turbo-Screw Design

The Turbo-Screw is typically used in foam extrusion processes as well as in processing heat-sensitive materials such as PLA. It is also known as Turbo-Cool screw. It is a new concept in screw flight design, increases total extrusion rate by the use of multiple cross-flight channels along the length of the flights. The Turbo-Screw design relies upon a series of new flight geometries, which promote cross-channel flow. Turbo-Screw comes in two configurations—(1) "C" type geometry and (2) "B" type geometry. Fig. 8.7a and b shows screw flight circulation channel geometries of two screw configurations [14,15].

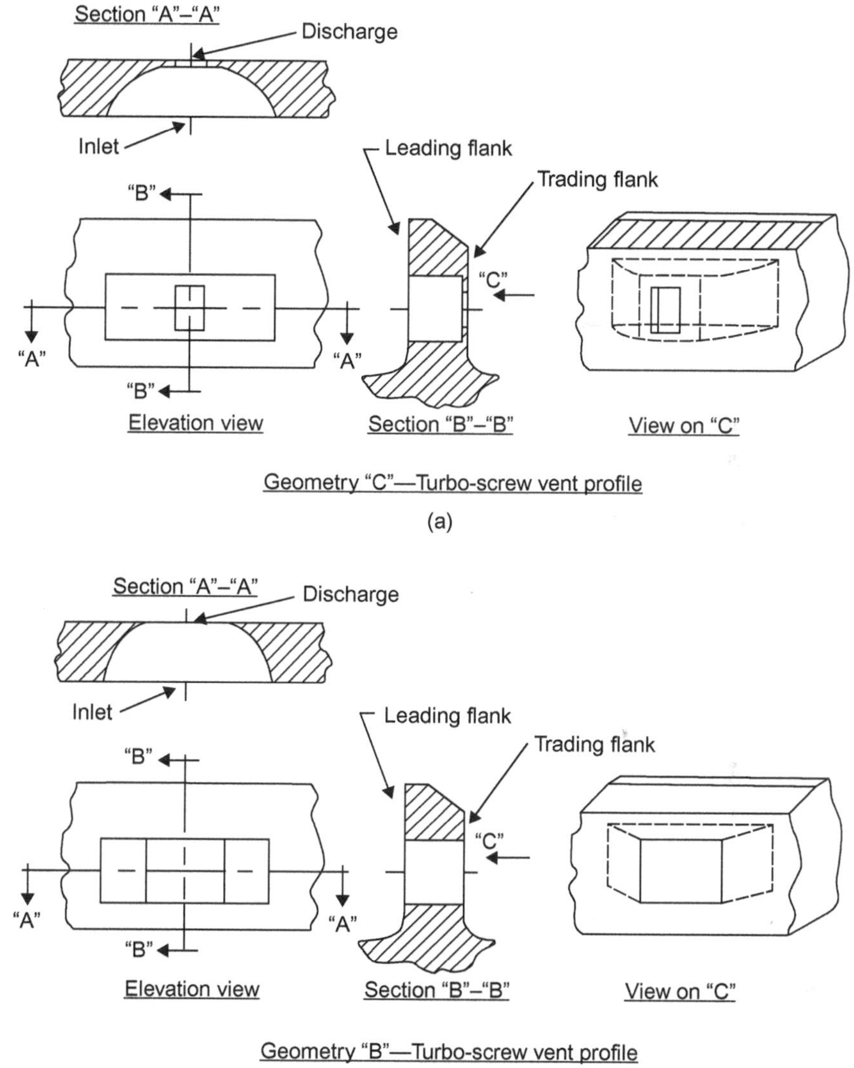

Figure 8.7 Screw flight circulation channel geometry (a) type "C" and (b) type "B" [14,15].

To minimize viscous heat generation in the channels, the Turbo-Screw is designed with a multiflight. Numerous openings on the flight cause the polymer melt to flow from the pushing side to the trailing side of the flight. This cross-flight flow results in efficient mixing and improved heat transfer. A perspective view of the Turbo-Screw is shown in Fig. 8.8 [14].

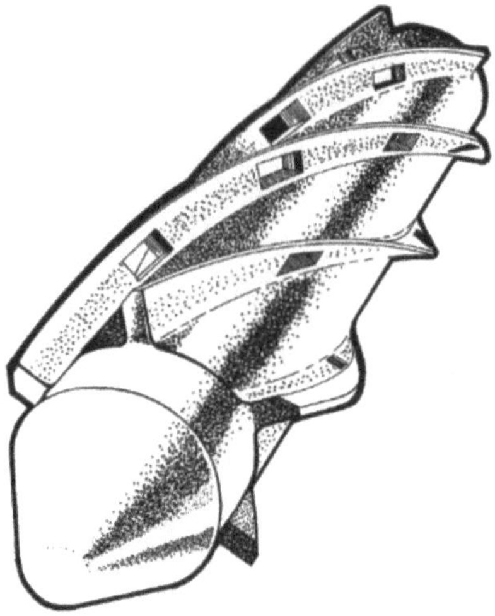

Figure 8.8 A perspective view of the Turbo-Screw [14].

As shown in Fig. 8.7a, the inlet of the flight circulation channel is on the leading flank of the flight. The discharge of the channel is labeled on view C-C in Fig. 8.7a. In order to promote elongational flow and vortices, the inlet of the flight circulation channel is 2 to 10 times as compared to the purposefully shaped discharge orifice. To achieve effective heat transfer from the polymer melt, it has been determined that the scooping effect of the curved sidewalls is critical [15].

As polymer melt flow through the flight circulation channel, it experiences an elongational flow. This stretches the polymer melt. As the melt leaves from the flight circulation channel, it experiences some degree of elastic recovery. This will tend to expand the leakage stream. In order to achieve efficient redistribution of the polymer melt, it is important that the acceleration of the polymer melt flows through the light circulation channel [15].

The inlet area of the flight circulation channel comprises 15% of the area of the pushing screw flank. The discharge area takes up about 4% of the area of the trailing screw flank. As a result, there is about a 4:1 reduction in cross-sectional area of the flight circulation channel. The geometry of the opposing wall pairs of the flight circulation channel is intended to promote mixing [15].

8.4.2 Pulverization Screw Design

Solid-state shear pulverization (SSSP) is a process that uses a modified TSE with cooling zones to maintain the polymer in the solid state

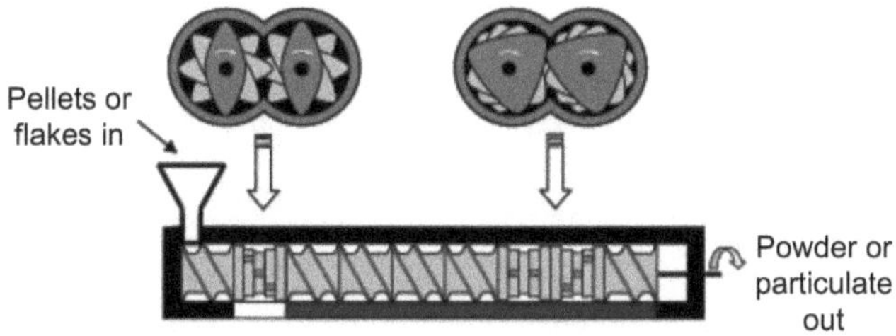

Figure 8.9 Schematics of solid-state shear pulverization extrusion [16].

during processing. High shear and compressive forces result in repeated fragmentation and fusion of polymer in the solid state producing excellent mixing and dispersion of polymer blends.

SSSP extrusion schematic is shown in Fig. 8.9 [16]. The SSSP setup is a modified TSE from Berstorff with the ability to cool the barrel below room temperature. Pulverization screw designs differ from conventional twin-screw extrusion due to the use of trilobe rather than bilobe screw elements along a portion of the screw.

Fig. 8.10 shows a patented SSSP extrusion setup. This setup uses a rotating conical meshing screw design to form fine powder by applying normal and shear forces on the polymeric material [17].

8.4.3 Reactive Extrusion Screw Design

Reactive extrusion refers to a continuous, chemical polymer modification or polymer synthesis. Reactive extrusions differ from conventional polymer modifications in that polymer analogue reactions can be performed without the need for solvents. Reactive extrusions also allow direct processing of the modified polymer, for example, blending with other polymers or producing a component using the injection molding process. Table 8.1 shows classification of candidates for reactive extrusion [18–21].

Reactive extrusion commonly utilize a two-stage screw design that comprises two compression sections separated by a decompression/extraction section, not too dissimilar to having two single-stage screws in series on a single shaft as shown in Fig. 8.11 [23]. Proper design of a two-stage screw requires understanding of the polymer viscosity, the output capacity of both stages, and the head pressure [23,24]. The two-stage screw first reduces the melt pressure to atmospheric to prevent polymer from flowing out the vent opening. This is accomplished by

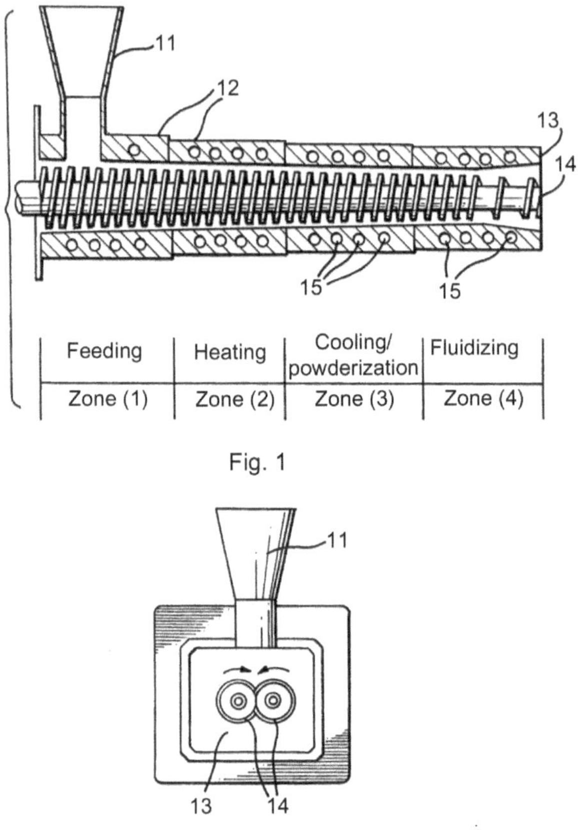

Figure 8.10 Solid-state shear extrusion pulverization patented setup [17].

increasing the channel volume so that the output of the extruder only partially fills the flights under the vent [25].

Reactive extrusion requires good distributive mixing of reactants while allowing sufficient residence time for the reaction to go to completion. This can be achieved by using co-rotating twin-screws. Application of alternating forward and reverse narrow kneading elements provides the highest degree of fill in the screw, with the longest average residence time and maximum longitudinal mixing. This screw arrangement results in plug flow transport and is used for polycondensation reaction to achieve narrow molecular weight distribution and high reaction rates. Screw configurations having only conveying elements are successfully used for ionic polymerizations in which back-mixing does not negatively influence the molecular weight distribution or the reaction rate [26].

Table 8.1 Classification of Candidates for Reactive Extrusion [22]

Type of reactive extrusion	Product obtained
Bulk polymerization	ε-caprolactone, PMMA, PU, PLA, PS, PA, PC, etc.
Graft copolymerization	MA-grafted PP, alkoxysilane-grafted PE, etc.
Reactive blending	Polymer blends, toughened PP, dynamic vulcanization, TPE, TPV, etc.
Chemical modification	Functional EVA, chain-extended PET, long-chain branched PP, etc.
Depolymerization	MW-modified PP, devulcanized rubber, etc.
Chemical recycling	L-lactide, styrene monomer, MMA monomer, etc.
Dewatering	Posttreated rubber, ABS, etc.

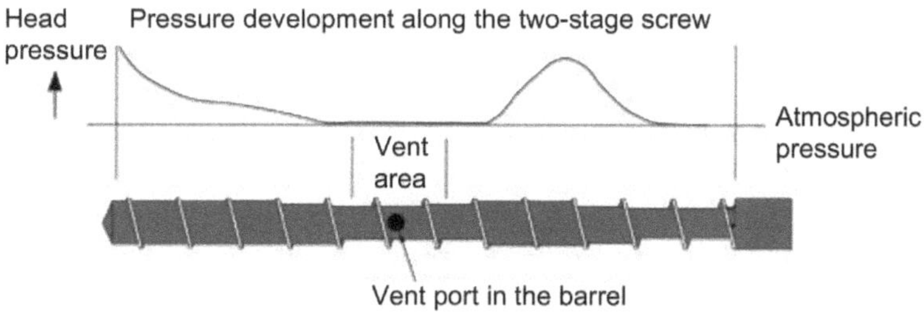

Figure 8.11 Two-stage screw design for reactive extrusion [23].

Vent openings are strategically located to remove residual monomer or reaction by-products. Stripping agents can be added to improve devolatilization efficiency with multiple vacuum zones [26].

References

[1] R.N.M. Agbisit, Relationship between material properties and microstructure—mechanical attributes of extruded biopolymeric foams (Doctoral dissertation), Kansas State University, 2007.

[2] Intelligent Control of Single Screw Polymer Extruder. <http://isac.wikidot.com/intelligent-control-of-single-screw-polymer-extruder>.

[3] Twin Screw Extruder Mixing Elements, <http://www.alibaba.com/product-detail/Nanjing-Kaiyou-HKY-65-plastic-twin_1645203468.html>.

[4] K. Luker, Single-screw extrusion and screw design, in: I. Ghebre-Selassie, C. Martin (Eds.), Pharmaceutical Extrusion Technology, Marcel Dekker, New York, 2003 (Chapter 3).

[5] C. Martin, Compounding PLA on twin-screws: what testing reveals, Plast. Technol. (2014)<http://www.ptonline.com/articles/compounding-pla-on-twin-screws-what-testing-reveals>.
[6] Words of wisdom: understanding twin-screw extruders: the high-speed energy input (HSEI) concept, Plast. Today. <http://www.plasticstoday.com/articles/words-wisdom-understanding-twin-screw-extruders-high-speed-energy-input-hsei-concept>, 2006.
[7] R. Steiner, Extruder design, in: I. Ghebre-Selassie, C. Martin (Eds.), Pharmaceutical Extrusion Technology, Marcel Dekker, New York, 2003, p. 26 (Chapter 2).
[8] G.J. Rokey, Single-screw extruders, in: M.N. Riaz (Ed.), Extruders in Food Applications, CRC Press, Boca Raton, FL, 2000, p. 25 (Chapter 2).
[9] G.R. Huber, Twin-screw extruders, in: M.N. Riaz (Ed.), Extruders in Food Applications, CRC Press, Boca Raton, FL, 2000, p. 81 (Chapter 5).
[10] F. Xie, P. Liu, L. Yu, Processing of plasticized starch-based materials, in: P. Halley, L. Averous (Eds.), Starch Polymers: From Genetic Engineering to Green Applications, Elsevier, Oxford, UK, 2014, p. 268 (Chapter 10).
[11] E. Chiellini, H. Gil, G. Braunegg, J. Buchert, P. Gatenholm, M.V.D. Zee, Biorelated Polymers: Sustainable Polymer Science and Technology, Springer, New York, 2013, p. 6.
[12] A.G.- Cunha, J.A. Covas, The Design of Extrusion Screws: An Optimization Approach, Dept. Polym. Eng., University of Minho, Portugal, 2001.
[13] H. Eslami, Understanding Screw Design for Film Extrusion Process, Macro Advanced Extrusion Systems. <http://www.macroeng.com/understanding-screw-design-for-film-extrusion-process.php>.
[14] J. Fogarty, D. Fogarty, C. Rauwendaal, A. Rios, Turbo-Screw™, new screw design for foam extrusion, Blowing Agents and Foaming Processes, Rapra Technology Ltd., Heidelberg, Germany, 2002.
[15] C. Rauwendaal, Polymer Mixing: A Self-Study Guide, Hanser Publishers, Munich, 1998.
[16] Novel Processing of Polymer Blends and Nanocomposites: Solid-State Shear Pulverization, Torkelson Research Group, Northwestern University. <http://torkelson.northwestern.edu/Research/SSSP/sssp.html>.
[17] F. Shutov, G. Ivanov, H. Arastoopour, Solid-state shear pulverization, European Patent EP 0596835 A1, Illinois Institute of Technology, 1994.
[18] Reactive Extrusion: Continuous Polymer Modification Using Co-Rotating Twin-Screw Extruders, Fraunhofer ICT. <http://www.ict.fraunhofer.de/content/dam/ict/en/documents/factsheets/PE_CE_Reaktivextrusion_E.pdf>.
[19] T. Sakai, Report on the state of the art: reactive processing using twin-screw extruders, Adv. Polymer Techn. 11 (2) (1991) 99–108.
[20] T. Sakai, N. Hashimoto, K. Kataoka, Experimental study on dewatering and devolatizing operations using intermeshed twin screw extruders, Int. Polym. Proc. 8 (3) (1993) 218.

[21] T. Fukushima, N. Hashimoto, Y. Sumihiro, T. Sakai, Y. Kimura, Development of a direct polycondensation process for Poly(L-Lactic acid), Int. Polym. Proc. 15 (4) (2000) 380–385.
[22] T. Sakai, Screw extrusion technology—past, present and future, Polymer 58 (2013) 11–12.
[23] J. Frankland, Extrusion know-how: the whys and hows (&ifs) of vented extruders, Plast. Technol. (January 2010)<http://www.ptonline.com/columns/extrusion-know-how-the-whys-hows-ifs-of-vented-extruders>.
[24] R. Hudson, Developments in the European Extrusion Industry: Machinery, Manufacturers, and Markets, iSmithers Rapra Publishing, Shawbury, UK, 1995.
[25] S.J. Derezinski, The vented barrier screw, in: Society of Plastics Engineers Annual Technical Conference, 2005.
[26] Y. Wang, Compounding in Co-rotating Twin-Screw Extruders, iSmithers Rapra Publishing, 2000, pp. 33–34.

9 Commercial Applications of Bioplastics

9.1 Introduction

Biodegradable plastics are the type of plastics that undergo decomposition over a period of time under composting conditions. The global biodegradable plastics market accounts for less than 1% of the overall plastics market, however, it is expected to grow at a fast pace over the next 5 years. The market growth is driven by continuous R&D activities, increased environmental awareness, and implementation of stringent environmental regulations. However, the high cost of biodegradable plastics has been a major barrier in the growth of the market [1].

The growth of biodegradable plastics is driven by the following factors:

1. Consumer preference toward environmental-friendly products
2. Use of renewable and bio-based raw material
3. Biodegradability
4. Government policies toward green procurement

Top players in the biodegradable plastics industry include Metabolix Inc. (U.S.), BASF SE (Germany), Corbion N.V. (Netherlands), NatureWorks LLC (U.S.), Biome Technology Plc (U.K.), and others [1].

The two most important commercial, biodegradable polymers are polylactic acid (PLA) and starch-based polymers, accounting for about 47% and 41%, respectively, of the total biodegradable polymer consumption. Starch sources vary worldwide, but include corn, potatoes, cassava, and sugar beets. In Europe, starch-based biodegradable polymers are the major type consumed, accounting for 62% of the market, due to Europe's large, starch-based capacity and their use in many applications. This is followed by PLA, with 24% and other biodegradable polymer types with 14%.

Biodegradable plastics have found their end-use in a variety of commercial applications. It is estimated that packaging and disposable housewares drive the biodegradable market. Foam packaging continues

Introduction to Bioplastics Engineering.
DOI: http://dx.doi.org/10.1016/B978-0-323-39396-6.00009-9

to represent significant market share for biodegradable plastics. Compostable bags, as well as single-use carrier plastic bags, follow foam packaging in terms of volume. Another developing use for biodegradable plastics is in the shale gas industry, where they are used during hydrofracking as more environmentally friendly proppants to "prop open" fractures in rock layers so oil and gas can be released [2].

Other areas where biodegradable plastics have seen their use are:

- Bags
- Agriculture and horticulture
- Disposable housewares
- Medical devices
- Consumer electronics
- Automotive

This chapter focuses on discussing different areas of commercial applications of biodegradable plastics such as packaging, bags and sacks, disposable housewares, agriculture and horticulture, medical devices, consumer electronics, and automotive.

9.2 Packaging

Biodegradable plastics, like conventional plastics, offer a large range of packaging applications. Some of the packaging options include bags for compost, agricultural foils, horticultures, nursery products, toys, and textiles. Other areas include packaging for contact articles including disposable cutlery, drinking cups, salad cups, plates, overwrap and lamination film, straws, stirrers, lids and cups, plates and containers. Fig. 9.1 shows a range of packaging options made from biodegradable plastics [3–9].

It is expected that the bioplastic materials used in food packaging applications perform by protecting the food from the environment and maintaining the food quality [10]. These protecting functions are obvious as it is important to be able to control and modify their mechanical and barrier properties which depend on the structure of the biopolymer. It is also important to study the change in properties of the packaging material during the length of time it was in contact with the food [11].

Fig. 9.2 shows two grades of biodegradable resins from Solaplast [12,13]. Both, Solaplast1723 and Solaplast1223 have found their use in the packaging industry. These two biopolymers that are derived from

Figure 9.1 Range of packaging options made from biodegradable plastics [3–9].

Figure 9.2 Solaplast resins are used extensively to manufacture biodegradable packaging [13].

algae are used for diverse applications. They can be bio-based, biodegradable, or durable. Each application depends on the type of biopolymer used. Packaging made from these biopolymers can be processed with all customary plastics processing technologies. No special machinery is required. Depending on the type of bioplastics used, only the processing parameters have to be adjusted [12].

9.3 Bags

Biodegradable shopping bags are made of polymers that degrade, or decompose, when exposed to air, water, or sunlight. Currently, CU Dining Services is promoting their use of biodegradable bags from Grab-n-Go's [14].

There are three main types of biodegradable bags:

Type 1: The original biodegradable bags, still found today, are made from resins containing starches, polyethylene, and heavy metals such as cadmium, lead, and beryllium.

Type 2: A second type has been invented using starches combined with biodegradable polymers such as PLA or BASF

EcoFlex. These bags meet ASTM compostable standards, while others do not [15].

Type 3: Oxo-Biodegradable bags use Totally Degradable Plastics Additives (TDPA™) to stimulate the breakdown of polymers and thus speed up the biodegradation process of conventional plastics [16,17].

Biodegradable bag labeled as a no. 4 (type 1) carries no information on meeting ASTM standards. Testing of the EPI Oxo-Biodegradable bags (type 3), as reported by the Biodegradable Products Institute, has found presence of moisture or humidity which inhibits onset of oxidation, which means that composting slows down the degradation of these bags [18]. Fig. 9.3 shows examples of different types of biodegradable bags [15−17,19,20].

A number of companies are developing bags made of biodegradable polymers for use in municipal and institutional composting programs. Such products would have to be truly compostable and conform to ASTM standard D6400−99; under the right environmental conditions, a compostable polymer breaks down into nothing more than carbon dioxide, water, a humus-like organic material, and some innocuous trace inorganics. Composting bags for yard waste or food scraps would also have to satisfy the operational requirements of composting facilities.

9.4 Disposable Housewares

Disposable housewares made from biodegradable plastics are now seen as marketable options, replacing traditional plastics such as polystyrene and polyolefin. Housewares such as kitchen tools and utensils, washable storage containers and cups, bathroom accessories, toys, hangers, and hooks are now being produced using biodegradable plastics. One such example is the Fantastic Beach toys by Zoë b Organic which are made with Mirel bioplastic from corn as shown in Fig. 9.4 [21].

Cutlery made from Cereplast, a biodegradable resin made from corn and potato starch, qualifies as a compostable plastic [22] and is marketed under the brand name Nat-Ur. NaturBag markets Natur-Ware cutlery which is both bio-based and certified compostable shown in Fig. 9.5 [23].

Fig. 9.6 shows biodegradable hangers from United Colors of Benetton which are 100% recyclable and biodegradable [24].

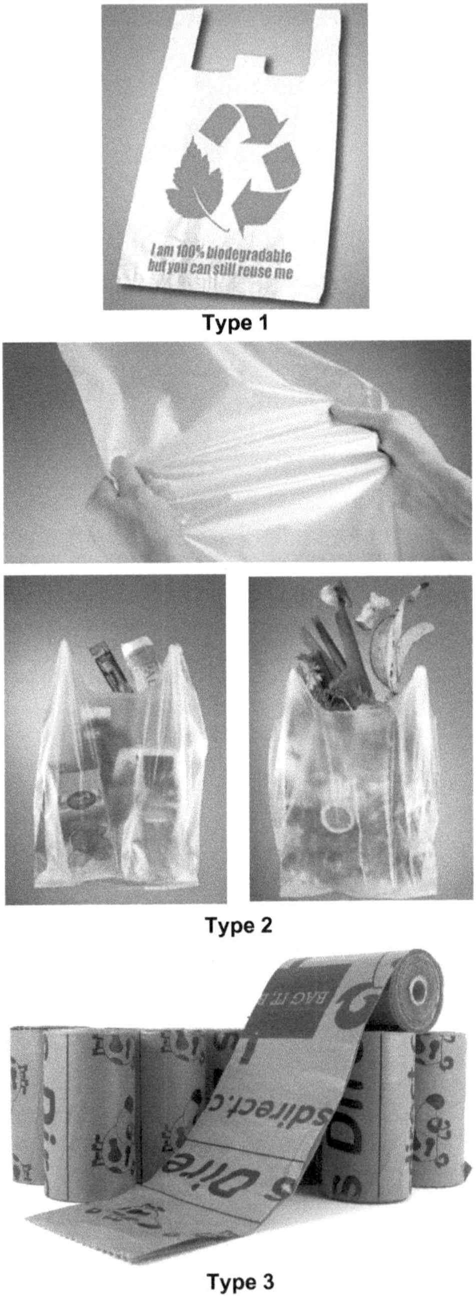

Figure 9.3 Different types of biodegradable bags [15–17,19,20].

Figure 9.4 Disposable housewares made from biodegradable plastics [21].

Figure 9.5 Natur-Ware bio-based and compostable cutlery [23].

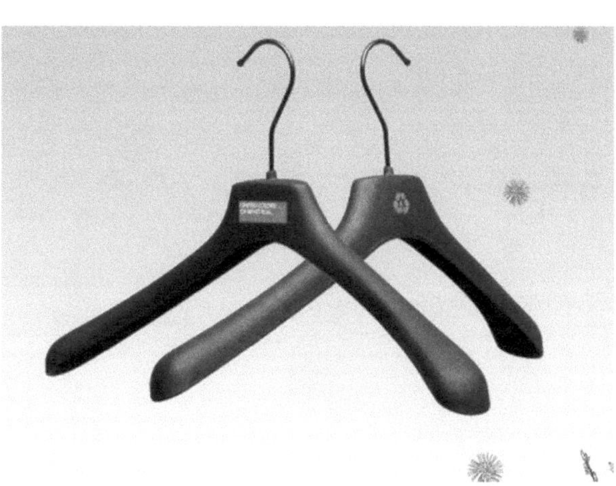

Figure 9.6 Biodegradable hangers from United Colors of Benetton [24].

9.5 Agriculture and Horticulture

Production of agricultural mulches, seeding strips, and tapes made from biodegradable polymers shown in Fig. 9.7 is another strong potential market [25,26]. The tape biodegrades in the soil as the seeds germinate and take root. Plastic mulch films are then used to give the new seedlings a head start in the spring; the mulch helps reduce evaporation and conserve moisture, increases soil temperature, and keeps control on the weeds. Compostable seed belts and active component capsulations made out of bioplastics have also proven to be beneficial.

Foils and nets made from degradable bioplastics are used in farming of mushrooms and the coating of tree and bush-roots. Foils, yarns, and nets made out of bioplastics help to secure freshly created slopes and mounds and protect them from erosion until the roots of the plants have developed sufficiently.

Solaplast mulching films are generally making rapid advances in this area. Solaplast 2100 series resins applications in agriculture and

Figure 9.7 Agricultural mulches, seeding strips, and tapes made from biodegradable plastic [25,26].

horticulture include films for banana bushes and grapevine bushes which have to be protected from dust and environmental influences. Fig. 9.8 shows Solaplast films for agriculture and horticulture sectors [27–29]. These films offer opportunities for marketing pot-planting. An example is potting herbs. The herb can simply be planted into their pot. Once the herbs are harvested, everything including the film can be composted. Some of the other applications of Solaplast 1223 may be for durable applications such as: fastening technology; plant pots for propagation/cultivation; fertilizer rods; or pheromone traps.

Fig. 9.9a shows plant baskets and pot made out of biodegradable materials. They are commonly used in cemeteries because of their biodegradable nature, convenience, and cost-efficiency. Operators of golf courses can benefit from the use of bioplastics by using degradable bioplastic tees as shown in Fig. 9.9b. This makes the collection of the remaining tees an obsolete practice; the tees simply remain in the ground to decompose [30,31].

9.6 Medical Devices

Nontoxic biodegradable polymer sutures are being used each year by surgeons in life-saving heart operations and other procedures. Easily sterilized, the sutures remain strong and intact until the surrounding tissues have healed. The sutures dissolve and are readily metabolized in the body leaving no trace.

Both pliable braided sutures as well as wiry monofilament sutures are available, depending on: (a) the type of surgery, (b) the knots to be used, (c) the tensile strength needed, and (d) the potential for infection. Fig. 9.10a shows biodegradable plastic staples which are used to close wounds and incisions. Plastic sutures are largely made from lactic or glycolic acid and account for about 95% of the total market, with various pins, implants, and dental devices making up much of the remainder. Fig. 9.10b–d shows biodegradable plastic pins, tacks, and screws which are used to hold shattered bones together while they heal, to reattach ligaments, and for delicate reconstructive surgery on ankles, knees, and hands [32–34].

Biodegradable dental implants, made of porous polymer particles, are being used to quickly fill the hole after a tooth has been extracted. Fig. 9.11 shows films of guided-tissue-regeneration membranes [35]. These membranes made from biodegradable polymer are used by dental

Figure 9.8 Solaplast compostable films [27–29].

(a) (b)

Figure 9.9 (a) Biodegradable burial pods and (b) PLA golf tee [30,31].

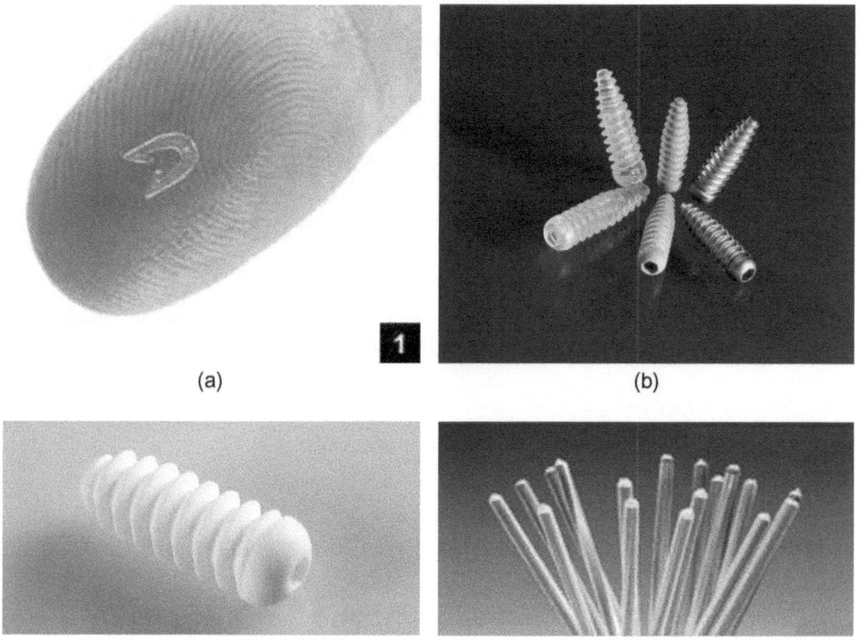

(c) (d)

Figure 9.10 (a) Insorb absorbable skin stapler, (b) dissolvable medical screws promoting bone growth, (c) absorbable anterior and posterior cruciate ligament interference screw, and (d) absorbable orthopedic pin [32–34].

Figure 9.11 Guided-tissue-regeneration membranes [35].

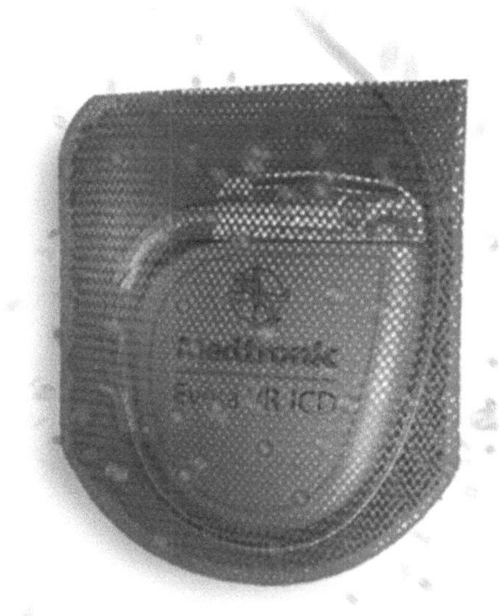

Figure 9.12 TYRX absorbable antibacterial envelope [37].

surgeons, to prevent scabbing and allow the slower-growing connective and ligament cells to regrow.

Fig. 9.12 shows TYRX absorbable antibacterial envelope, a biodegradable tyrosine-derived polyarylate which is a critical component of a new hernia repair device marketed by TyRx Pharma, Inc. This firm focuses on the development of new drug-eluting medical devices. Developed as a coating on a surgical mesh, these polymers improve the handling capabilities of the mesh to facilitate precise placement during surgery. The coating is then reabsorbed, leaving a smaller amount of material in the body [36].

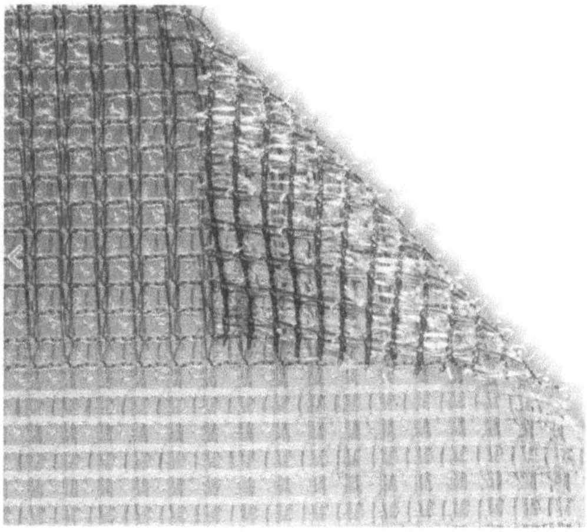

Figure 9.13 Parietix ProGrip™ laparoscopic self-fixating mesh [38].

Fig. 9.13 shows Parietex ProGrip™ laparoscopic self-fixating mesh that facilitates tension-free repair. It is the first bicomponent mesh composed of monofilament polyester and a resorbable PLA gripping system. The mesh's self-gripping feature provides secure fixation, reducing, or possibly eliminating the need to place supplemental fixation. This mesh is made from lightweight, hydrophilic polyester and provides optimal porosity which helps in better tissue growth and reduction in pain. In addition, it also reduces the amount of leftover foreign material in the body [38].

Two novel classes of biodegradable polymers, PolyActive and OctoDEX are two drug delivery systems that are licensed out by OctoPlus. These biodegradable systems are used for the site-specific delivery of drugs. Because of their biodegradability and linear release properties, these two systems are an excellent technology for the controlled release of proteins and lipophilic small molecules for both local and systemic administration, and have applications in pharmaceuticals and medical technology [39].

Molecular structure of PolyActive is shown in Fig. 9.14. It represents a series of poly (ether ester) multiblock copolymers, based on polyethylene glycol (PEG) and PBT. Major advantage of this system is the ability to vary the amount and length of each of the two building blocks, creating a diverse family of customized polymers. Polymer matrix characteristics such as rate of controlled release, degradation, swelling, and strength can be precisely controlled by the appropriate combination of the two copolymer segments [39,40].

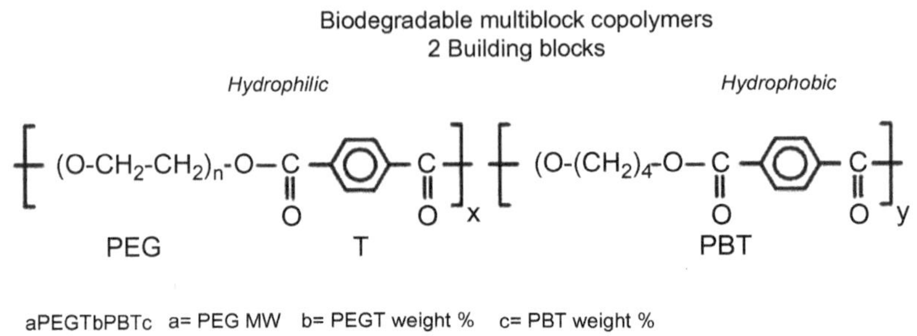

Figure 9.14 Molecular structure of proprietary controlled release technology PolyActive [40].

9.7 Consumer Electronics

A large proportion of today's consumer market is made from plastics. In electronics market, plastics are used in casings, circuit boards, and data storage because of their durability, toughness, lightness, and mobility. However, there has been a surge of bioplastic products which are being introduced in the fast-moving consumer electronics sector such as touch screen computer casings, loud speakers, keyboard elements, mobile casings, vacuum cleaners, and a mouse for a laptop.

SUPLA has developed optimized PLA compounds for the consumer electronics industry based on lactides from Corbion Purac. The launching application is the world's first bioplastic touch screen computer that is developed in collaboration with a Taiwanese OEM/ODM, Kuender, as shown in Fig. 9.15. The high gloss housing of this computer is made from high heat PLA [41–44].

The PLA blends used for the monitor screens also bring improved impact resistance, excellent high gloss finish, and stable, precise processing. Some other examples of biodegradable consumer electronics products include computer keyboards, ear phones, mobile phones, laptops, games consoles, and tablets. Some of the products are shown in Fig. 9.16 [41].

9.8 Automotive

Automotive industry is going through a major change with the focus on reducing fuel consumption and emissions by reducing a vehicle's weight. Plastics have been a part of automotive components but there

Figure 9.15 Bioplastic touch screen computer casing developed in cooperation with leading Taiwanese OEM/ODM of consumer electronics—Kuender [41–44].

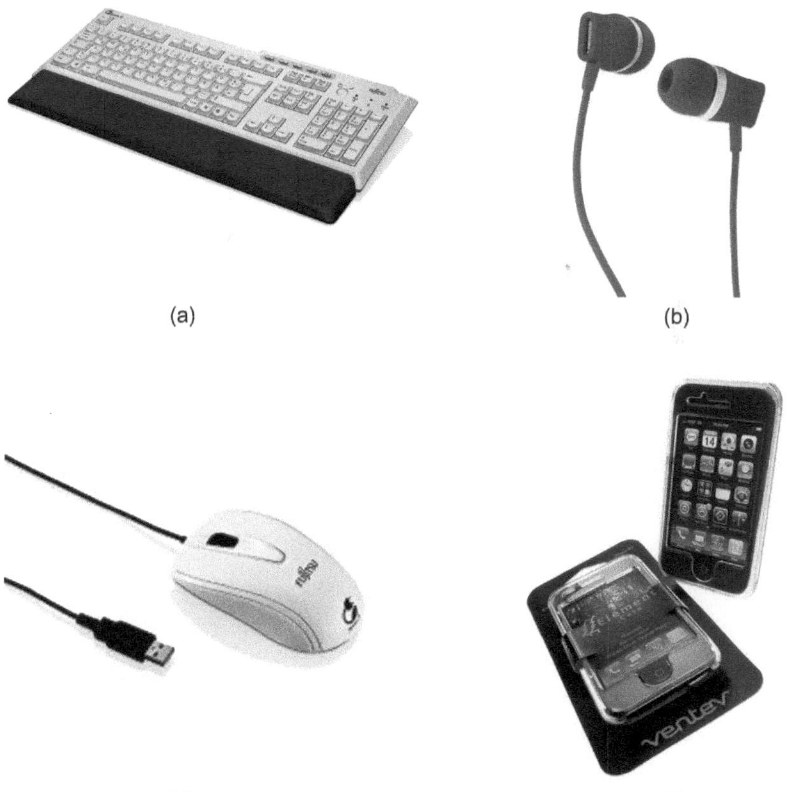

Figure 9.16 Consumer electronics products—(a) Fujitsu's® new computer keyboard, (b) EOps Noisezero i + Eco edition using cornstarch bioplastics, (c) Fujitsu's® biodegradable mouse, (d) Ventev™ element case made from 100% Naturacell™, (e) biodegradable iPhone case realized with Apinat, (f) drum cover of Fuji Xerox copy machine, (g) solar charger Xindao sunshine, and (h) XD design Ginkgo solar tree [41].

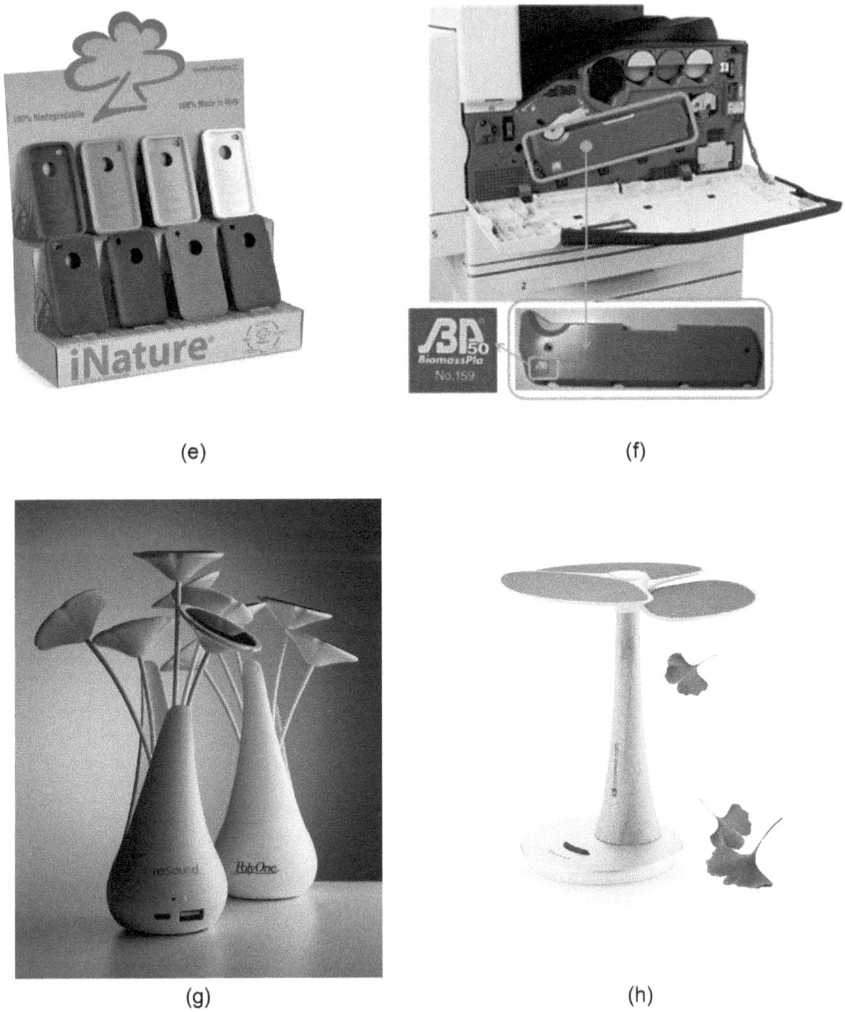

Figure 9.16 Continued

has been a concern toward plastics recyclability. For this reason, automotive industry is making a big leap toward using bioplastics components. Some of the bio-based plastics such as bio-based polyamides and bio-based polyesters are already successfully being used by leading automotive brands around the world today with the aim of reducing their products' environmental impact. Besides these evident advantages, for example bio-based performance polymers (bio-based polyesters and polypropylene/bio-based polyamides, bio-based poly propylene)

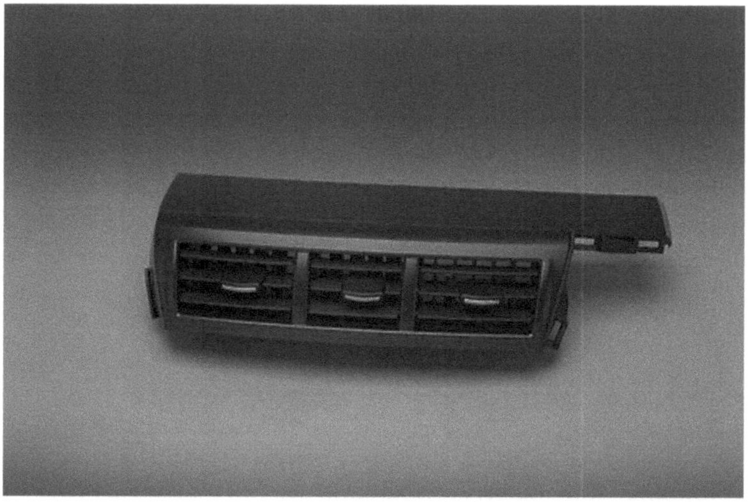

Figure 9.17 Toyota Prius vent louver of Sorona EP PTT polyester [46].

feature all the performance criteria important for high-quality car components [45].

Bioplastics have reached maturity as a suitable material for a large number of automotive applications, offering high performance and a unique potential for reducing a product's environmental impact. Manufacturers have turned to bio-based or partly bio-based durable bioplastics to produce sturdy dashboard components as well as solid interior and exterior features. Components made completely or partially from bioplastics can provide a standard of safety that is of ultimate importance in the transportation sector. The products include seat and airbag covers as well as steering wheels.

The world's first commercial automotive application for DuPont's Sorona EP PTT polyester is interior vent louver vanes for the Toyota Prius shown in Fig. 9.17 [46]. Sorona EP has 37% renewable content derived from fermentation of corn sugar. Sorona 2045G with 45% glass was chosen for its combination of surface appearance and mechanical properties—environmental benefits were not the primary driver. PTT allowed design of thinner vanes because it is stronger and stiffer than PBT or nylon 6 at similar glass loadings [45].

A front-runner in adopting bioplastics is Japanese car manufacturer Toyota which uses bioplastics such as bio-based polyesters, bio-based PET, and PLA-blends in its production process. Toyota has set the target of replacing 20% by weight of all oil-based plastics for cars with bioplastics by 2015. The Toyota SAI and Toyota Prius models already feature a number of bioplastic applications such as headliners, sun visors, or floor mats. Up to 60% of the interior fabrics are made of bio-based polyesters, which provide mechanical properties equal to or even better than PBT [47].

Compared to PBT bio-based polyesters [45,47]:

- provide a higher stiffness
- feature very good dimensional stability and lower warpage
- feature better thermal shock resistance
- provide good electrical properties
- are easier to process
- provide a better surface gloss

Toyota has typically used bio-based polypropylene/polylactic acid (PP/PLA) composite derived from plant material. They have developed a new plant-derived bio-based plastic more suitable for auto interiors than other bio-based plastics. Fig. 9.18a shows the list of vehicles containing bio-based plastics. Toyota began using the new material in the luggage compartment liner of the new Lexus CT200h hybrid-electric compact car. This is the world's first use of a bio-based polyethylene terephthalate (PET) resin in the auto industry. Fig. 9.18b shows an example of eco-plastic usage in Lexus HS 250h [48].

Fig. 9.19 shows an automotive air filter box and interior trim parts supplied by Röchling automotive [49]. These parts are produced using high heat PLA compounds, based on lactides from Corbion Purac, under the name of Plantura. Plantura is a material family based on PLA, suitable for automotive applications as well as for durable goods. These materials are said to show improved hydrolysis and thermal

MODEL	PART
Prius	Seat cushions, scuff plate, cowl side trim, register blades
Corolla	Seat cushions
Matrix	Seat cushions
RAV4	Seat cushions
Lexus RX 350	Seat cushions
Lexus HS 250h	Luggage trim upholstery, cowl side trim, door scuff plate, tool box area, floor finish plate, seat cushions, package tray
Lexus CT 200h	Luggage compartment liner, carpeting
Camry	Radiator end-tank

(a)

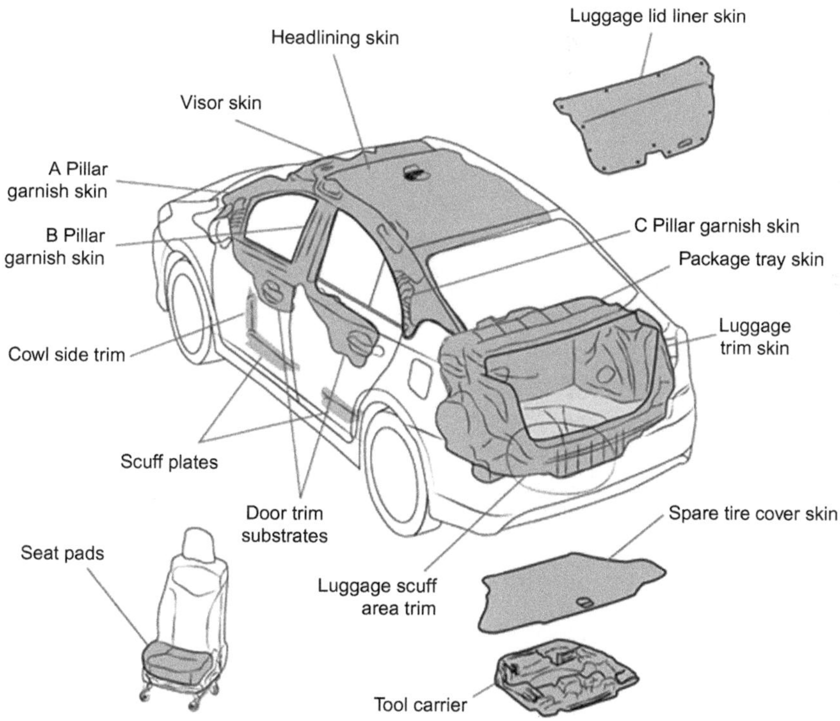

(b)

Figure 9.18 Vehicles containing bio-based plastics [48].

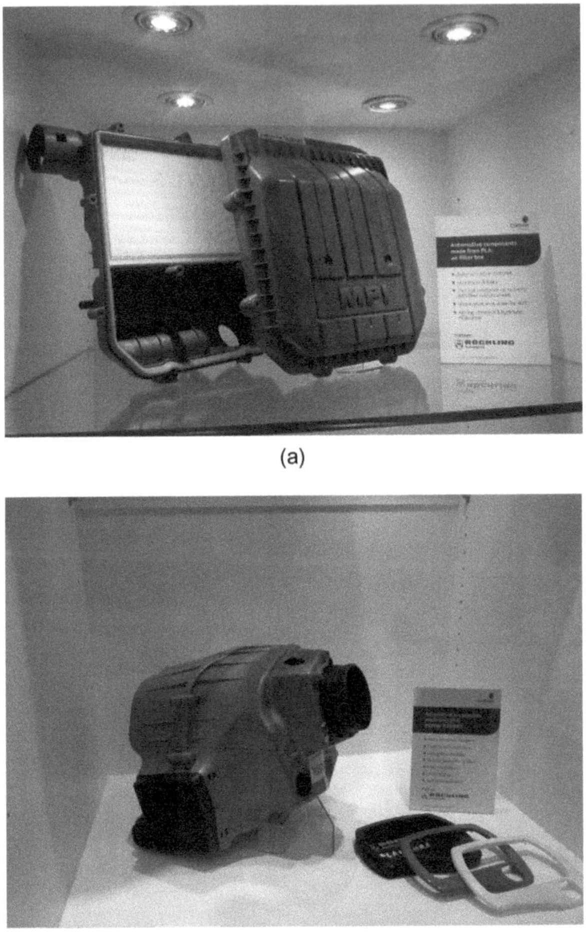

(a)

(b)

Figure 9.19 PLA used for (a) air filter box and (b) interior trim supplied by Röchling automotive [49].

resistance up to 140°C with fiber reinforcement as well as a good scratch and UV resistance [49].

References

[1] Plastics Business Magazine, <http://www.plasticsbusinessmag.com/enews/stories/050713/biodegradable.shtml#.VjgZZZca6Ao>.

[2] Markets and Markets, Biodegradable plastics market by type and by application—Global trends & forecasts to 2020, <http://www.marketsandmarkets.com/Market-Reports/biodegradable-plastics-93.html>.

[3] Papstar Products, <http://www.papstar-products.com/Disposable-tableware-and-service-packaging/Disposable-tableware-and-articles-made-of-renewable-raw-materials-biodegradabl/Disposable-articles-made-of-wood-and-bamboo/Cutlery/100-Forks-wood-pure-16–5-cm.htm?shop = papstar_pe&a = article&ProdNr = 18199&t = 100&c = 100300&p = 100300>.

[4] Bio Packing, PSM Cutlery, <http://www.biopacking.com/Product-Cornstarch%20Cutlery.html>.

[5] Biodegradable Bendy Drinking Straws, Envirogadget, <http://www.envirogadget.com/odd-gadgets/biodegradable-bendy-drinking-straws/>.

[6] Biodegradable Plastic Cups, Health and Safety Signs, <http://www.health-safety-signs.uk.com/cgi/products.pl?environmentally = sustainable_products&id = 13>.

[7] French Wine, <http://www.french-wine-web.com/wines/swizzle-sticks/1000-count-7-long-wooden-coffee-stirrers-biodegradable-earth-friendly-stir-wood-skewers-craft-popsicle-cupcake-stick-pack-of-1000-pcs>.

[8] Compostable Packaging Test: Bambu Plates Breakdown, Inhabitat, <http://inhabitat.com/compostable-packaging-test-bambu-plates-breakdown/>.

[9] Top 4 Consumer Behavior Trends That Should Influence Your Packaging, <http://blog.customboxesnow.com/category/recycling-environmentally-friendly-boxes/>.

[10] I.S. Arvanitoyannis, Totally and partially biodegradable polymer blends based on natural synthetic macromolecules: preparation, physical properties, and potential as food packaging materials, J. Macromol. Sci. Rev. Macromol. Chem. Phys. C39 (2) (1999) 205–271.

[11] G. Scott, Green polymer, Polym. Degrada. Stab. 68 (2000) 1–7.

[12] Solaplast, Target Markets and Applications, <http://algix.com/solaplast-applications/>.

[13] Solaplast, Solaplast Resin Grades, <http://algix.com/products-services/solaplast-resin-grades/>.

[14] Housing and Dining Services, University of Colorado Boulder, <https://housing.colorado.edu/dining/locations-hours/cu-run-grab-n-go>.

[15] R. Narayan, S. Mojo, Summary of ASTM D6400 Test Method and Specifications and Correlation of Test Methods to Real World Composting Results, International Biodegradable Products Institute, 2002 <http://www.bpiworld.org/BPI-Public/News/Article.html>.

[16] EPI Environmental Technologies Inc., EPI-Global, <http://www.epi-global.com/en/Index-e.htm>, 2005.

[17] TDPA®: Totally Degradable Plastic Additive, EPI Environmental Technologies Inc., <http://www.epi-global.com/en/about-tdpa.php>.

[18] S. Mojo, BPI Assessment of Oxo-Degradable Films, Biodegradable Products Institute, May, 2007.

[19] Plastics Portal, <http://www.plasticsportal.net/wa/plasticsEU~en_GB/function/conversions:/publish/common/upload/biodegradable_plastics/ecoflex_sense.pdf>.
[20] Biodegradable Plastic Bags Carry More Ecological Harm Than Good, <http://myzerowaste.com/2009/06/biodegradable-plastic-bags-carry-more-ecological-harm-than-good/>.
[21] E. Erschfeld, Biodegradable Beach Toys: Safe for Kids and Oceans, <http://www.organicspamagazine.com/biodegradable-beach-toys-safe-for-kids-and-oceans/>.
[22] J. Thomas, Compostable cutlery made from Cereplast, <http://www.treehugger.com/clean-technology/compostable-cutlery-made-from-cereplast.html>.
[23] Bio-based compostable cutlery, NaturBag, <http://www.naturbag.com/biobased-compostable-cutlery>.
[24] J.M. Chua, Benetton gets greener with biodegradable hangers and organic cotton, <http://www.ecouterre.com/benetton-gets-greener-with-biodegradable-hangers-organic-cotton-garments/>.
[25] Agriculture and horticulture, European Bioplastics, <http://en.european-bioplastics.org/market/applications/agriculturehorticulture>.
[26] A.M. Mohan, Miracle-gro compostable seed pods are consumer-focused, <http://www.greenerpackage.com/compost_biodegrade/miracle-gro_compostable_seed_pods_are_consumer-focused>.
[27] Target markets and applications, Algix®, <http://algix.com/solaplast-applications/>.
[28] Image gallery, Algix®, <http://algix.com/image-gallery/#!>.
[29] Bio-plastics.org, <http://www.bio-plastics.org/en/information--knowledge-a-market-know-how/range-of-applications/farming-and-gardening>.
[30] A.N. Desai, Burial pods: Biodegradable coffins could one day replace depressing cemeteries with lush green forests, <http://www.inquisitr.com/1888594/burial-pods-biodegradable-coffins-could-one-day-replace-depressing-cemeteries-with-lush-green-forests/>.
[31] Shenzhen Bio Plastic Technology, <http://www.bio-bp.com/e_productshow/?27-PLA-Golf-Tee-27.html>.
[32] S. Fox, A new breed of medical screws dissolve in body and promote bone growth, <http://www.popsci.com/science/article/2010-03/biodegradable-medical-screws-disolve-body-and-promote-bone-growth>.
[33] Euroscrew® TCP, Teknimed, <http://www.medicalexpo.com/prod/teknimed/product-70291-499892.html>.
[34] Product review—Insorb absorbable skin stapler, <http://exploreplasticsurgery.com/category/medical-devices/>.
[35] Atelocollagen membranes Hypro-Sorb F, Guided tissue regeneration (GTR), <http://www.hypro.cz/en-membrany.html>.
[36] V. Rai, The academic journey of Dr. Joachim Kohn, professor at Rutgers University, Controlled Release Society, <http://www.controlledreleasesociety.org/publications/intrack/Pages/it0012.aspx>.

[37] TYRX™ absorbable antibacterial envelope, Medtronic Corp., <http://tyrx.com/products/tyrx-absorbable/index.htm>.

[38] ProGrip™ Laparoscopic Self-Fixating Mesh, Medtronic Corp., <http://www.medtronic.com/covidien/products/hernia-repair/progrip-laparoscopic-self-fixating-mesh>.

[39] PolyActive Microspheres, OctoPlus, <http://www.octoplus.nl/nl/Expertise/PolyActive%C2%AE-microspheres>.

[40] PolyActive™ Factsheet, <http://www.afinitica.com/arnews/sites/default/files/techdocs/PolyActive_Nov20061%5B1%5D.pdf>.

[41] Consumer electronics images, European Bioplastics Magazine, <http://en.european-bioplastics.org/press/press-pictures/consumer-electronics/>.

[42] Solutions to reducing consumer electronic waste with biodegradable polymers, Brigit, <http://www.brigit-project.eu/detalle_noticia.php?no_id =2618>.

[43] D.D. Guzman, Corbion Purac highlights new PLA uses, Green Chemical Blog, <http://greenchemicalsblog.com/2013/10/16/corbion-purac-highlights-new-pla-uses/>.

[44] S. Moore, PLA bioplastic debuts in touchscreen computer housing, Plastics Today, <http://www.plasticstoday.com/articles/k-2013-pla-bio-plastic-debuts-touchscreen-computer-housing>, October 2013.

[45] Sorona® EP Thermoplastic Polymer, Product datasheet, DuPont, <www2.dupont.com/Plastics/en_US/Products/Sorona/sorona_ep.html>.

[46] Bio-based resins score in auto applications, Plastics Technology, <http://www.ptonline.com/articles/biobased-resins-score-in-auto-applications>, January 2012.

[47] Bioplastics in the automotive market—Clear benefits and strong performance, Fact Sheet, European Bioplastics Magazine, January 2015.

[48] Green Innovation, Toyota Corp., <http://www.toyota.com/usa/environmentreport2012/green_innovation.html>.

[49] Latest generation of PLA, Corbion Purac, <http://www.plastics.gl/exhibit/latest-generation-of-pla/>.

10 New Developments

10.1 Introduction

Bioplastics are made wholly or in part from renewable biomass sources such as sugarcane and corn, or from microbe such as yeast. Some bioplastics are biodegradable or even compostable, under the right conditions. Bioplastics made from renewable resources can be naturally recycled by biological processes, thus limiting the use of fossil fuels and protecting the environment.

Therefore, bioplastics are sustainable, largely biodegradable, and biocompatible. Today, bioplastics have become a necessity in many industrial applications such as food packaging, agriculture and horticulture, composting bags, and hygiene. Bioplastics have also found their use in biomedical, structural, electrical, and other consumer products. Packaging has been, and still is, one of the most common uses of bioplastics, but there is growth in other areas, such as textiles and automotive applications.

With increasing demand for global plastic consumption, a lot of research is being dedicated toward exploring green materials and new ways to process them. This chapter focuses on new developments of bioplastics in material, processing, and applications.

10.2 Materials

There has been a marked increase in the use of biodegradable materials for a range of applications. Some of the applications include:

- Compostable bags
- Packaging
- Agriculture and horticulture
- Medical devices
- Consumer electronics
- Automotive

Introduction to Bioplastics Engineering.
DOI: http://dx.doi.org/10.1016/B978-0-323-39396-6.00010-5

Among all the applications, it is estimated that packaging and disposable housewares drive the biodegradable market. There are several new plastics that have been launched that are biodegradable.

A newly developed family of biodegradable polymers has shown potential for use in intracellular delivery and sustained release of therapeutic drugs to the acidic environments of tumors, inflammatory tissues, and intracellular vesicles that hold foreign matter. These polymers are called polyketals. Polyketals are biodegradable and FDA approved with added advantages over existing biodegradable polymers. A generalized structure of polyketal is shown in Fig. 10.1 [1].

Synthesis is a simple and easily customized process. Degradation of the polymer does not produce inflammation-causing acid, but instead generates membrane-permeable products that allow all of the polymer's by-products to diffuse outside the cell. That means that by-products shouldn't accumulate in a patient's tissue and cause inflammation. However, polyketal nanoparticles use the cell's acid to hydrolyze into hydrophilic compounds that can release encapsulated therapeutics at an accelerated rate in the acidic environments to which they are targeted. Also, unlike polyester-based biomaterials, polyketal nanoparticles do not generate acid when they degrade [1].

In drug delivery applications, where most of the illnesses being treated require weekly injections, polyketals hydrolyze in a week, diffuse out of the cell, and are then excreted outside of the cell. On the other hand, polyesters take months to degrade which make them unsuitable for drug delivery applications. Another potential application of polyketals includes the delivery of antioxidants to treat acute liver failure in people who have suffered an alcohol or acetaminophen overdose. In these patients, the liver stops functioning because macrophage cells in the liver create reactive oxygen species. One of the treatments is the delivery of superoxide dismutase, an enzyme that essentially detoxifies superoxide. Polyketals are also used in any type of protein-based vaccine and protein delivery for a wide range of therapeutics, including insulin delivery for Type 1 diabetics [1].

Rolco Inc. of Kasota, MN launched a line of board game pieces made of thermoplastic biocomposite compounds that contain up to 50%

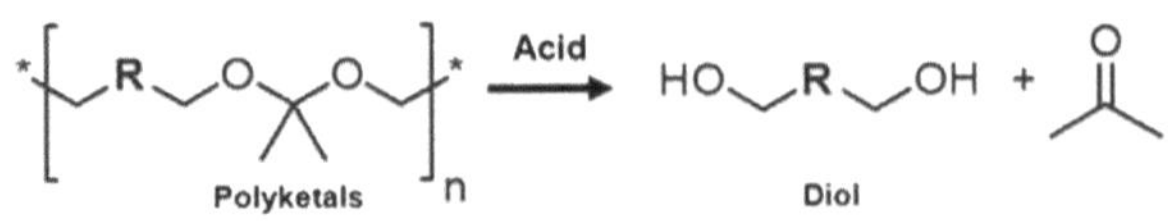

Figure 10.1 Structure of polyketals [1].

recycled wood by-product as shown in Fig. 10.2. The compound combines agricultural waste, recycled plastics, and patented additives. The waste includes material such as wood flour or rice hulls and alloys them with recycled polymers using proprietary, patented additives. The resulting material combines the durability of plastic with the appearance and workability of wood [2].

ArmorLite, a US company has developed a new roofing material that is 100% recyclable and produces zero waste during manufacturing [3]. ArmorLite developed this revolutionary new product in collaboration with General Electric, using advanced engineering polymer materials. The top layer of the roofing product is made from a highly weatherable material called GE Geloy for superior weatherability. An example of roofing material from ArmorLite is shown in Fig. 10.3 [4].

Figure 10.2 Board game pieces made of thermoplastic biocomposite compounds containing 50% recycled wood by-product [2].

Figure 10.3 An example of roofing material from ArmorLite [4].

The ArmorLite roofing material is made from an acrylic-styrene-acrylonitrile (ASA) and makes them unique because of three main reasons:

1. Reflects heat back into atmosphere and provides insulation
2. 100% recyclable with zero waste in manufacturing
3. Engineered to withstand the most extreme climatic conditions

Polyester-amide (PEA) polymers used for specific drug delivery are widely recognized as naturally biodegradable and biocompatible compounds which naturally degrade in the body into amino acid components, the building blocks of proteins. SurModics offers five biodegradable polymers suitable for site-specific drug delivery—PolyActive and OctoDEX licensed by Octoplus, SynBiosys licensed by InnoCore, and Eureka, which is internally developed. PEA polymers are well suited for delivering small molecule drugs [5].

Perstorp, a Swedish specialty chemical company has developed polycaprolactone (PCL)-based bioplastics is and marketed under the brand name Capa. This biodegradable plastic is highly miscible and blends easily with other plastics, melts easily, and is nontoxic. Other biopolymers can be added with Capa to improve overall mechanical properties, biodegradability, and compatibility across a wide range of polymers [6].

Fig. 10.4 shows Capa brand PCL used as film. Capa improves the functional properties, especially toughness and flexibility, of bioplastics

Figure 10.4 Capa brand polycaprolactone (PCL) used as film [6].

such as polylactic acid (PLA), polyhydroxyalkanoate (PHA), and starch, enabling these to be competitive in film and packaging applications. Its compostability offers end-of-life benefits in packaging, bags, and film applications [6].

Sulzer Chemtech has developed an innovative process that allows the production of new grades of PLA with heat stability up to 180°C [7]. Sulzer uses a continuous process with two reactors arranged in series followed by a devolatilization unit. The first reactor is a Sulzer Mixer Reactor (SMR) which runs in a loop configuration. The second reactor is a straight-through SMR reactor. Both reactors have static mixers inside and run at relatively high temperatures, heated by the catalyst reaction. Polymerization occurs along the inside shell of the SMR cylinder. The static mixers inside the reactors are a series of bent tubes through which a cooling medium travels to remove excess heat and control the reaction rate of polymerization [8].

The devolatilization unit strips off any unpolymerized monomer and the polymer then goes directly into a pelletizer. Currently, Sulzer's PLA process is only available for use with Purac's lactides. Lactide is shipped as an inert flake in dry bulk containers. It is critical that lactides are transported and stored in a way that it is protected from moisture by special packaging, or it reverts to lactic acid again [8].

Researchers in Belgium have developed a simpler, more efficient, and economic way to process and produce PLA bioplastic. In this technique, researchers borrowed a petrochemical concept that uses a zeolite as a catalyst in the reactor to guide the chemical process that converts lactic acid into lactide. Zeolite is a porous mineral composed of aluminum, silicon, and oxygen. By selecting a specific type of zeolite on the basis of its pore shape, they were able to convert lactic acid directly into the building blocks for PLA without making the larger by-products that do not fit into the zeolite pores [9].

10.2.1 Expanded PLA (EPLA) Bead and Foams

Expanded polystyrene (EPS) made from fossil-based polymers has been extensively applied to different applications. More awareness to protect the environment has resulted in a shift to move toward bio-based products. This move has inspired a push to develop biofoams as a replacement for EPS. PLA is one such promising biologically derived alternative which can be used for a range of commodity plastics.

Biopolymer Network has a patent application for a new method to produce E-PLA biofoams from commercially available PLA beads using a novel carbon dioxide (CO_2) foaming technology. Table 10.1 shows comparison of properties of E-PLA to EPS. This technology can be used to produce both foamed beads and molded shapes and products such as insulation, packaging, and specialty components. This technology has been used to produce molded E-PLA product on standard commercial polystyrene molding equipment and tests on these products have shown similar performance to polystyrene for both impact resistance and insulation [10].

Fig. 10.5 shows a schematic of E-PLA manufacturing [10]. Using Biopolymer Network's novel carbon dioxide (CO_2) foaming technology, the commercial PLA pellets are impregnated. This step is followed by temperature conditioning of the pellets. These pellets can be stored in

Table 10.1 Properties Comparison of E-PLA to EPS [10]

Laboratory tests have shown the following properties of E-PLA compared to EPS					
	EPS	**E-PLA**			
Density (g/L)	25	25	35	60	100
Thermal resistance (m^2.K/W)	~0.68	~0.69			
Thermal conductivity (W/m.k)	~0.035	~0.035			
Compressive modulus (MPa)	5.3	2.9	5.3	11.3	20.9
Compressive strength (MPa)	0.12	0.07	0.12	0.25	0.45
Compressive stress at 10% strain (MPa)	0.16	0.10	0.16	0.33	0.60
Shear modulus (MPa)	3.2	3.0	3.7	5.4	8.0
Shear strength (MPa)	0.12	0.11	0.15	0.25	0.41
Shear stress, maximum (MPa)	0.17	0.18	0.23	0.36	0.55

Standard test methods: Compression ASTM D1621-00; Shear ASTM C273-00.

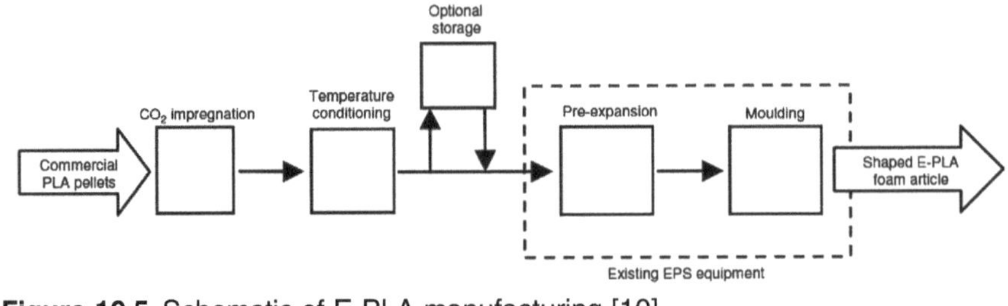

Figure 10.5 Schematic of E-PLA manufacturing [10].

moisture-free air tight containers or used directly for production. This process uses existing EPS equipment [10].

A range of products that Biopolymer Network made from E-PLA are shown in Fig. 10.6 [11,12].

Synbra Technology bv., the raw material development and supply unit of Synbra Group acquired the first license for a new PLA reactor technology developed by Purac in the Netherlands and Sulzer Chemtech AG in Switzerland. Both, Purac and Sulzer Chemtech, produce lactide monomer which is the precursor of PLA, and aid in providing Synbra with all the capabilities to produce PLA products [8].

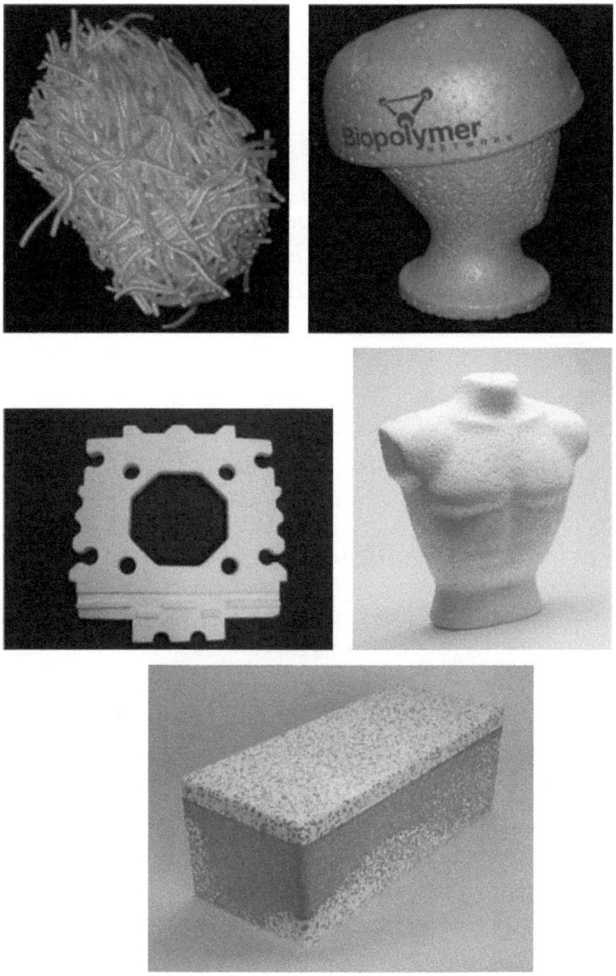

Figure 10.6 Examples of expanded PLA products [11,12].

Figure 10.7 Example of BioFoam [13].

Synbra's EPLA is trademarked BioFoam, and is colored light green to distinguish it from conventional white EPS. BioFoam is aimed at specialty packaging for consumer goods and at bio-derived cushion filling. In the Netherlands, the market leader in beanbag furniture, Fatboy the Original bv., has signed up to use BioFoam beads for filling. An example of BioFoam is shown in Fig. 10.7 [13].

Top-load strength of EPLA is comparable to that of conventional EPS without the problems associated with pentane blowing agents. EPLA beads contain no VOCs and have no shelf-life limitation. Synbra already plans to bring BioFoam to North America [13].

10.2.2 Ecodear Nanoalloys

Ecodear Nanoalloy features a lineup of products for use in injection molding and extrusion processes by leveraging Toray's Nanoalloy technology to blend PLA with other materials to achieve a good balance of mechanical properties and processability.

Toray's Nanoalloy technology is an innovative microstructure control technology developed by Toray that can bring about dramatic improvement in characteristics compared to existing materials by minutely dispersing multiple polymers on a nano-metric scale. The technology enables high performance and functionality in polymers which could not be achieved in conventional alloys of micron-metric scale.

As shown in Fig. 10.8, Nanoalloy technology enables nano-level dispersion of resins with differing ratios of hard and soft components to

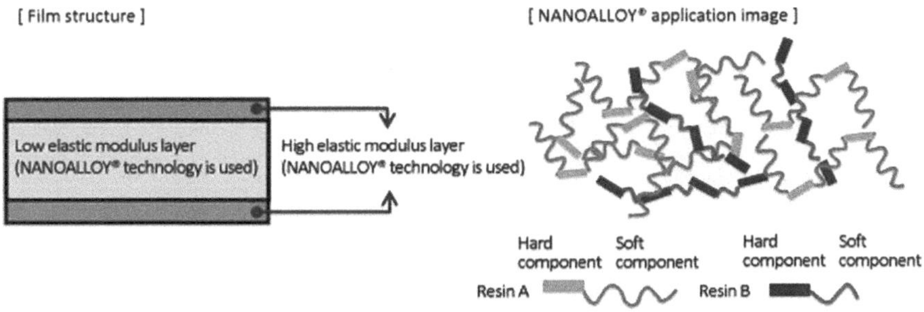

Figure 10.8 Effects of Nanoalloy technology on formability [14].

freely control elastic modulus at a given temperature. Furthermore, Toray developed a highly formable film by realizing finer control of elastic modulus at a given temperature through highly accurate lamination of layers with varying elastic modulus using the high precision lamination technology in film forming [14].

Some of the new values created by Nanoalloy technology are shown in Table 10.2 [15].

Fig. 10.9 shows some of the typical products made from Toray's Ecodear Nanoalloy brand of resins [16–18].

10.2.3 Ecovio L Foam

Ecovio L Foam is a refinement of Ecovio LBX 8145, the first plastic made of renewable raw materials, which BASF introduced onto the market in early 2006 as the umbrella brand for all future biodegradable compounds based on renewable raw materials [19].

Like Ecovio L Foam, Ecovio LBX also consists of Ecoflex, BASF's biodegradable plastic on a petrochemical basis, and PLA, which is made of corn, a renewable raw material. Like with the predecessor product Ecovio LBX, the main benefit for users is that they can utilize the material as a ready-made product, that is to say, directly and without further blending or preparation. Ecovio LBX lends itself particularly well for the extrusion of biodegradable films. It can even be modified in such a way that it is suitable for injection-molding and deep-drawing applications [20].

10.2.4 Ingeo 3D Series

Ingeo 3D850 is a grade developed for manufacturing 3D printer monofilament. This grade exhibits faster crystallization rates and is able

Table 10.2 Properties of Nanoalloy Technology [15]

Properties	Applications	New value
Shock resistance	Golf clubs (shafts), tennis racket frames and strings, bicycles, fishing rods, bats, sports protectors, automobile exteriors, safety helmets, mattress, and flooring materials	Improves the flexibility under high velocity impact as well as shock absorbing and restoration properties of polymers
High flowability	Small precision parts such as telecommunication equipment connectors	Improves the flowability of polymer in the melt state, and not only enables thinner parts and product design with complex shapes but also is expected to achieve energy savings through shorter molding cycles and lower molding temperatures while reducing greenhouse gases
Formability	Decorative films for smart devices, home electronic appliances, and automobile interior as well as exterior	Form complex shapes at low molding temperatures. Replaces existing technologies such as painting and plating; it enables decorations on 3D designs using environmental-friendly films
Vibration absorption	Transportation equipment, building materials, factory equipment, and machine tools	Enables creation of plastic materials that reduce vibration and noise
Heat resistance	Base films for data backup tapes, electrical insulating materials	Improves heat resistance and service life

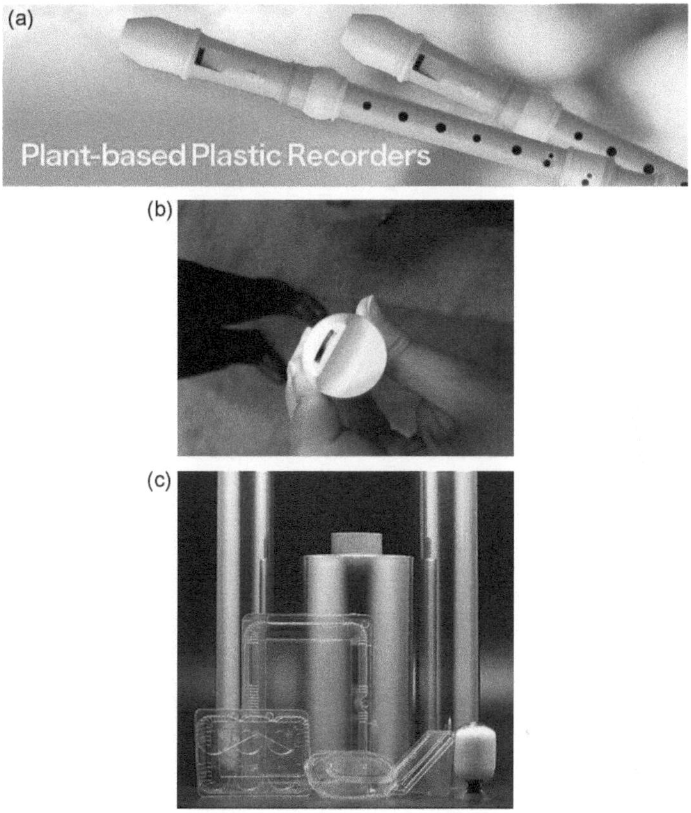

Figure 10.9 Ecodear Nanoalloy in (a) plant-based plastics recorder, (b) Yamaha 400 series soprano recorder, and (c) ECODEAR Polylactic Acid (PLA) films [16–18].

to develop improved heat-resistance in 3D printed parts [21]. This low color resin grade demonstrates the best performance in formulated systems designed to enhance toughness or heat-resistance. Monofilaments made with Ingeo 3D850 have excellent 3D printing characteristics such as precise detail, good adhesion to build plates, less warping or curling, and low odor [22]. PLA filament made from Ingeo demonstrates excellent properties in 3D printing applications as shown in Table 10.3 [23].

10.2.5 Expansorb

Expansorb consists of a range of compatible and biodegradable polymers, which include poly (D, L- lactide) (PLA), and poly (D, L-lactide, co-glycolide acid) (PLGA). PLA/PLGA biopolymers are primarily used

Table 10.3 Comparison between Ingeo 3D Biopolymer and Conventional Plastic [23]

Performance requirement	Market expectation	Ingeo performance	ABS performance
Temperature in printing	Lower is better for safety and control	190–230°C possible, sharp melting behavior	230–250°C
Vapor or odor in printing	Low or no emissions, smell small particulate	Very low emissions and lower particulate count	Strong negative odor and higher particulate count
Heated build chamber	Ideally not required and IP limits implementation for some users	Does not require heated build chamber, but opportunity to leverage	Required to avoid warping while manufacturing larger parts
Fusing performance	Fuses easily, sticks to self and build chamber	Good fusing and build platform adhesion	Difficult to stick to certain build platforms
Thermal shrinkage	Low shrinkage	Excellent, low shrink	Warps on larger builds

Figure 10.10 Expansorb-B universal expander liquid [25].

for controlled and sustained drug delivery system of drug substances. These polymers are used to overcome multiple injections, high toxicity, and to achieve the desired therapeutic effect. The polymer protects the molecules with short in-vivo half-lives [24]. The application includes the controlled parenteral administration or depot of drugs such as Peptides, therapeutic proteins, hormones, antibiotics, etc. An example is shown in Fig. 10.10 [25].

10.2.6 LACTEL Absorbable Polymers

Durect currently designs, develops, and manufactures a range of biodegradable polymers based on lactide, glycolide, and caprolactone under the LACTEL brand for pharmaceutical and medical device applications as raw materials [26]. An example of LACTEL absorbable polymer is shown in Fig. 10.11 [27].

The polymers and copolymers of glycolide, lactide, and ε-caprolactone degrade by simple hydrolysis. When exposed to aqueous media such as buffers or tissue, water is absorbed and reacts with the ester linkages, thus breaking the polymer backbone. One hydroxyl group and one carboxylic acid group form for each ester linkage that is hydrolyzed. Because the ester linkages are cleaved randomly along the polymer backbone, relatively few water-soluble fragments form

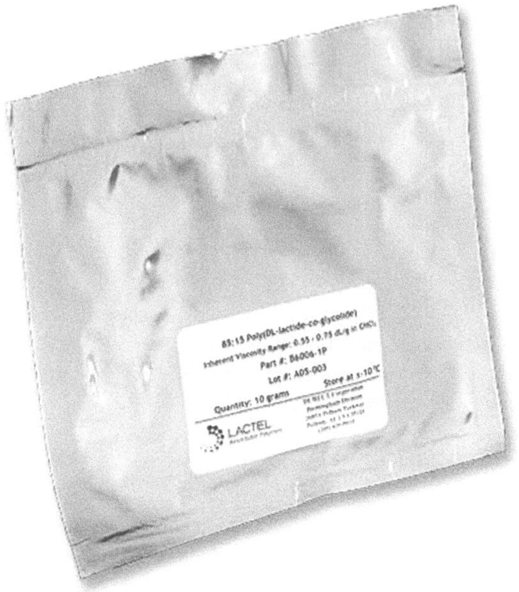

Figure 10.11 LACTEL absorbable polymer [27].

initially. Rather, long polymer chains are broken to form shorter ones. As a result, the reduction in molecular weight produces an increase in hydrophilicity but little change initially in physical properties or polymer mass [26].

As degradation proceeds, however, the reduction in molecular weight leads to a reduction in physical properties and the formation of water-soluble fragments. These water-soluble fragments diffuse away from the polymer and are ultimately hydrolyzed to glycolic and lactic acids, which are processed through normal metabolic pathways. The rate of hydrolysis depends on factors such as the size and hydrophilicity of the particular polymer implant, the monomer composition and degree of crystallinity of the polymer, and the pH and temperature of the environment [26].

10.3 Processing

US injection molders are still pretty green when it comes to processing the new crop of renewably sourced biopolymers. These biologically derived polymers made from PLA, PHA, and starch-based resins

Figure 10.12 Recycled chicken feathers can create a new class of biodegradable plastics [28].

are attracting growing market interest as materials with no ties to petrochemical-based thermoplastics.

Institute of Agriculture & Natural Resources at University of Nebraska-Lincoln have developed plastics from renewable resources to replace those derived from petroleum products. Recycled chicken features can create a new class of biodegradable plastics (Fig. 10.12) [28]. The new water-resistant thermoplastic, processed chicken feathers with chemicals, including methyl acrylate, a colorless liquid found in nail polish that undergoes polymerization. This process is used to produce plastics in which molecules link together one by one into huge chains and result in films which are defined as feather-*g*-poly(methyl acrylate) plastic.

This renewable plastic had excellent properties as a thermoplastic, was substantially stronger and more resistant to tearing than plastics made from soy protein or starch, and as a first among chicken-feather plastics had good resistance to water [28].

Patented technology developed at the University of Massachusetts Lowell for melt blending PLA biopolymer with 20% of any of several biodegradable aliphatic copolyesters (US Pat. 5883199) is said to achieve dramatic improvements in mechanical properties over PLA alone. Elongation can be 200% versus about 7% for plain PLA, according to the University of Massachusetts researchers. The blends reportedly retain their properties after aging, while PLA alone rapidly gets brittle. Such results were obtained with blends of PLA plus polycaprolactone, BASF's Ecoflex, Showa Hipolymer's Bionolle, or Metabolix's Mirel PHA polyhydroxyalkanoate [29].

10.4 Applications

The variety of bioplastics application is steadily expanding that offers high performance, new properties, and aids in cutting the use of resources.

The German company QMILK presented its textile fiber and bioplastic based on the milk protein casein which is considered unsuitable for use with food products (Fig. 10.13). QMILK is based on 100% renewable resources and can be freely modified and copolymerized [30].

Fischerwerke developed a product FIS Green 300T, the first bio-based injection mortar with application targeting construction industry as shown in Fig. 10.14 [32]. These chemical injection mortars are

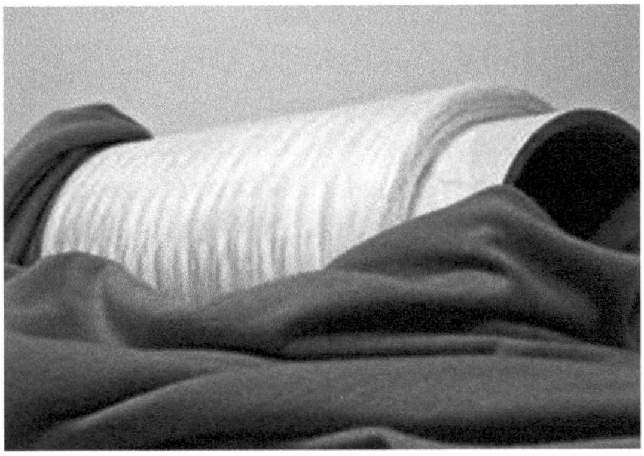

Figure 10.13 QMILK's textile fiber bioplastic-based on milk protein casein [30].

Figure 10.14 Fischerweke's first bio-based injection mortar, FIS Green 300T [31].

complex material mixtures. This product contains bio-based carbon content of 50–85 % achieved and certified by DIN CERTCO [31].

The French company Arkema, introduced a ski boot made from the new rigid bio-based polyamide block-based elastomer Pebax RNew 80R53. This ski boot known as Scarpa shown in Fig. 10.15 contains more than 90% of bio-based content. It is also lightweight, has long-term UV resistance, and increased rigidity. The product offers multiple options for advanced boot design [30].

The Austrian company Witasek developed a compostable tree protector BioWit-tube as shown in Fig. 10.16 [30,32]. The product is made of

Figure 10.15 Arkema's new ski boot, Scarpa made from rigid bio-based polyamide block-based elastomer Pebax RNew 80R53 [30].

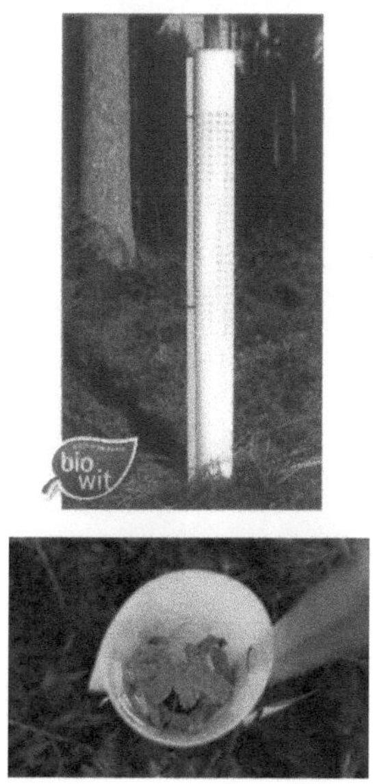

Figure 10.16 Witasek's compostable tree protector BioWit tube [30,32].

a transparent film that ensures healthy plant growth by protecting it against game browsing and fraying. The tube features a new locking system, which enables easy transport, and saves on freight and storage costs [32].

ZinCo GmbH developed system substrates made from recycled and pure-grade clay. These system substrates are used in drainage elements and protection mats. Zinco's natureline product lines made from PLA and bio-based polyethylene (Bio-PE) are used in filtration, drainage, and protection mats as shown in Fig. 10.17 [33].

ZIEHL-ABEGG's Bionic Bio-Fan made from castor oil-based polyamide is shown in Fig. 10.18 [34]. This product is fully recyclable, cut

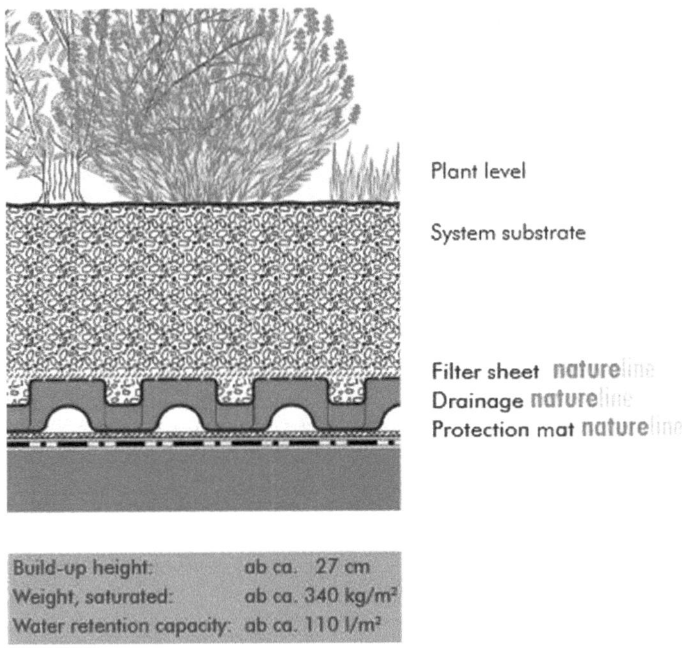

Figure 10.17 ZinCo's system substrate made from renewable resources [33].

Figure 10.18 ZIEHL-ABEGG's Bionic Bio-Fan [34].

down on noise, electricity use, and carbon dioxide (CO_2) emissions. The Bio-Fan is 6% lighter in comparison with products made of fossil-based polyamide (PA6) GF30 primarily due to 5% reduction in density and lower moisture absorption (1.2%) [30,34].

In Finland, consumers are buying milk in cartons, made by Tetra Pak. Tetra Pak has developed milk cartons that are 100% plant-based. Fig. 10.19a shows an example of the different kinds of milk cartons. A small company called Iris Industries used Kickstarter to get off the ground with Denimite, a marbleized blue composite made of recycled denim and a thermoset resin binding agent that is partially bio-based as shown in Fig. 10.19b. Fig. 10.19c shows an example of Mushroom packaging developed by Ecovative. This packing material also called as Myco Foam is designed to replace polystyrene packaging used for take-out meals and beverage [35–38].

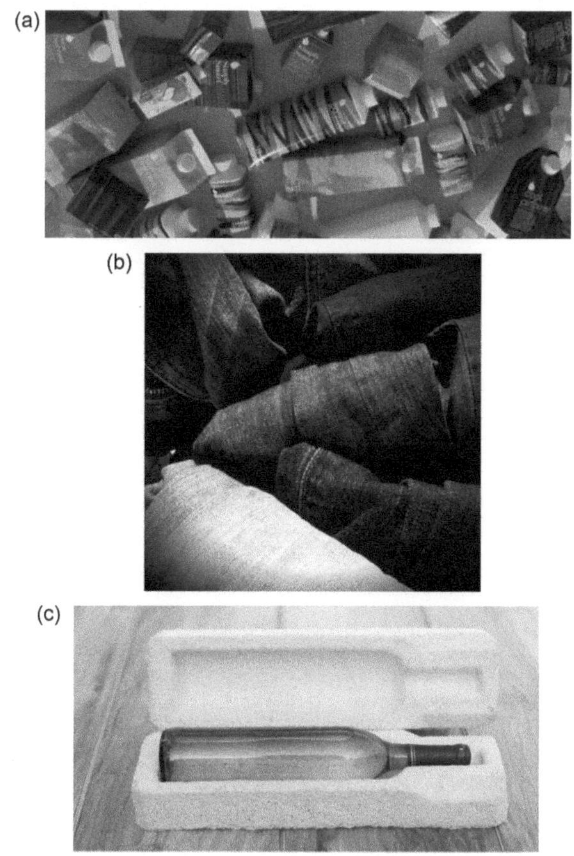

Figure 10.19 (a) Tetra Pak milk cartons, (b) Denimite product, and (c) Ecovative Myco foam packaging [35–38].

Because the demand is sharply increasing for food packaging made from biodegradable materials, BASF has introduced Ecovio L foam for the packaging industry. Typical products that use these products are the fast-food boxes and food-product trays as shown in Fig. 10.20 [19].

NatureWorks Ingeo shrink and laminated films are made from plant unlike the usual wrap, films, and laminates which are made from oil-based materials [39]. These films provide performance with eco-friendly benefits. They are suitable for bakery, shrink wraps, laminated coatings, envelope windows, and multilayered performance packaging [40]. Fig. 10.21 shows a debuted yogurt with a shrink sleeve made from Ingeo [39].

Figure 10.20 Food boxes made from Ecovio L foam [19].

Figure 10.21 Debuted Yogurt 7.0, a pro-biotic yogurt with EarthFirst shrink sleeve label made from Ingeo [39].

Vancouver-based G-Kup Coffee received a patent [41] for what it claims is the world's first 100% biodegradable and compostable single-serving coffee pod. As an alternative, G-Kups are held together with a bamboo and sugar cane sleeve, with a biodegradable polymer lining that can withstand boiling water [42]. An example of G-Kup coffee pods are shown in Fig. 10.22 [42].

Figure 10.22 G-Kup Coffee's 100% compostable coffee pods [42].

Figure 10.23 Front panel of Sharp's new smartphone made from Durabio [43].

Mitsubishi Chemical Corp. developed bio-based engineering plastics, Durabio made from plant-derived isosorbide. Durabio is a high-performance material that offers greater resistance to impact, heat, and weather than conventional engineering plastics. In addition, it has excellent transparency and low optical distortion. Jointly together with Sharp Corp., they developed the front panel of Sharp's new smartphone as shown in Fig. 10.23 [43].

References

[1] Polyketals, A new biomaterial for drug delivery, <http://groups.bme.gatech.edu/groups/murthylab/murthylab/polyketals.html>.

[2] Rolco Games, <http://www.rolcogames.com/pages/home/1>.

[3] R. Weinberg, AmorLitet roofing technology revolutionizes archaic industry with greenest product ever conceived, <http://www.ewire.com/newsreleases/armorlite–roofing-technology-revolutionizes-archaic-industrywith-greenest-product-ever-conceived/>.

[4] Transforming roof tile, shake and slate with SABIC innovative plastics weatherable GELOY™ XTW resin, <https://www.sabic-ip.com/gep/en/LiteratureAndTools/CaseStudyDetail/armorliteroofing.html>.

[5] P. Ankeny, SurModics, Inc., SurModics gain access to two biodegradable polymer families from OctoPlus for medical device applications, <http://www.businesswire.com/news/home/20040629005104/en/SurModics-Gains-Access-Biodegradable-Polymer-Families-OctoPlus>.

[6] R. Lingle, Two promising new developments in bioplastics for packaging, Sustainable Packaging, <http://www.packagingdigest.com/sustainable-packaging/2-promising-developments-in-bioplastics-for-packaging141110>, November 2014.

[7] Great advances in bioplastics, Sulzer Technical Review, Sulzer Chemtech, <http://www.sulzer.com/en/-/media/Documents/Cross_Division/STR/2012/STR_2012_1_13_14_Wintergerste_e.pdf>, January 2012.

[8] J.H. Schut, PLA Biopolymers: new copolymers, expandable beads, engineering alloys and more, Plastics Technol. (November 2008)<http://www.ptonline.com/products/pla-biopolymers-new-copolymers-expandable-beads-engineering-alloys-and-more>.

[9] A. Pasolini, New production process makes PLA bioplastic cheaper and greener, <http://www.gizmag.com/bioplastic-pla-cheaper-production-process/38498/>.

[10] Expanded Polylactic Acid (E-PLA): An eco-friendly alternative to EPS, Technical Bulletin, Biopolymer Network, <http://www.biopolymernetwork.com/Data/Default/Documents/E-PLA%20flyer.pdf>.

[11] E-PLA—ZealaFoam, Biopolymer Network, <http://www.biopolymernetwork.com/content/E-Pla-Foam/37.aspx>.

[12] ZealaFoam, Biopolymer Network, <http://www.biopolymernetwork.com/content/Zealafoam/48.aspx>.

[13] BioFoam®, Synbra Technology bv, <http://www.synbratechnology.nl/index.php?page=biofoam-4>.

[14] Nanoalloy Technology, Toray, <http://www.nanoalloy.jp/en/effect/eff_005.html>.

[15] Technical Impact, Nanoalloy Technology, Toray, <http://www.nanoalloy.jp/en/effect/index.html>.

[16] Plant-based plastic recorders, Yamaha, <http://usa.yamaha.com/products/musical-instruments/winds/recorders/plant-based_plastic/#tab=feature>.

[17] Yamaha 400 series soprano recorder, Tom Lee Music, <http://tomleemusic.ca/blog/yamaha-400-series-soprano-recorder/>.

[18] ECODEAR™ Polylactic Acid (PLA) Film, Plastics and Chemicals, Toray, <http://www.toray.com/products/films/fil_005.html>.

[19] F. Kruecki, The development of biodegradable and bio-based plastics goes on, News releases, BASF Chemicals, <http://www.plasticsportal.net/wa/plasticsEU~en_GB/portal/show/common/plasticsportal_news/2007/07_269>.

[20] Ecovio®, Transforming visions into realities, BASF Chemicals, <http://www.easyfairs.com/uploads/tx_ef/Ecovio_Brochure_min.pdf>.

[21] NatureWorks™ launches new initiative to support the growing 3D printing market, News and Events, NatureWorks™, <http://www.natureworksllc.com/News-and-Events/Press-Releases/2015/05-12-15-Ingeo-PLA-formulations-for-3D-Printing>.

[22] Ingeo™ Biopolymer 3D850 Technical Data Sheet, 3D Printing Monofilament—High Heat Grade, NatureWorks™, <http://www.natureworksllc.com/~/media/Technical_Resources/Technical_Data_Sheets/TechnicalDataSheet_3D850_Monofilament_pdf.pdf?la=en>.

[23] Ingeo™ PLA Filament in 3D Printing, Product and Applications, NatureWorks™, <http://www.natureworksllc.com/Product-and-Applications/3D-Printing>.

[24] Biodegradable Polymers (PLA/PLGA), Pharma Synthesis, PCAS, <http://www.pcas.com/expansorb-range-resorbable-polymer-excipient.html>.

[25] Expansorb®, Protechno, <http://www.protechno.com/index.php?id_product=227&controller=product&id_lang=4>.

[26] LACTEL® Absorbable Polymers, <http://www.absorbables.com/>.

[27] Packaging, Storage and Handling, LACTEL® Absorbable Polymers, <http://www.absorbables.com/products/storage.html>.

[28] Advance toward making biodegradable plastics from waste chicken feathers, American Chemical Society, News Releases, <http://www.acs.org/content/acs/en/pressroom/newsreleases/2011/march/advance-toward-making-biodegradable-plastics-from-waste-chicken-feathers.html>, March 2011.

[29] PLA/Biopolymer blends reduce brittleness, Plastics Technology, <http://www.ptonline.com/products/pla-biopolymer-blends-reduce-brittleness>, September 2007.
[30] 6 new bioplastics products—from clothes to construction, European Bioplastics Bulletin, <http://en.european-bioplastics.org/blog/2014/04/15/6-new-bioplastics-products-from-clothes-to-construction/>.
[31] Fischer FIS GREEN 300 T, Fischerweke, Biowerkstoff, <http://www.biowerkstoff-kongress.de/media/files/Award/14fischerFISGreen300T.pdf>.
[32] BioWit® Tube, Kressibucher, <http://www.kressibucher.ch/assets/preisliste/biowit-tube.pdf>.
[33] Natureline, ZinCo GmbH, <http://www.zinco-greenroof.com/EN/downloads/pdfs/ZinCo_Natureline.pdf>.
[34] Bionic Bio-Fan®, Ziehl-Abegg, Biowerkstoff, <http://www.biowerkstoff-kongress.de/media/files/Award/14ZIEHL-ABEGGSEBionicBio-Fan.pdf>.
[35] From tea bags to Miatas, bioplastics are on the rise, National Science Foundation, <http://www.nsf.gov/discoveries/disc_summ.jsp?cntn_id=134326>.
[36] Tetra Pak, <http://www.tetrapak.com/fi>.
[37] Recycled Denim, IRIS Industries, <http://www.iris-industries.com/recycled-denim/>.
[38] Myco Foam, Mushroom® Packaging, Ecovative, <http://www.ecovativedesign.com/myco-foam>.
[39] Ingeo™ News, NatureWorks™, 5 (1) (May 2008), <http://www.natureworksllc.com/News-and-Events/IngeoNews/IngeoNews-V5/IngeoNews-V5-Issue1>.
[40] Can a simple plastic film wrap really make a difference, NatureWorks™, <http://www.natureworksllc.com/Product-and-Applications/Films>.
[41] D.J. Footz, Biodegradable and compostable single-serve beverage ingredient package, US Pat. 20140335236 A1, November 2014.
[42] T. Dawes, Vancouver's G-Kup ready to disrupt massively wasteful single-serve coffee pod market, Cantech Lett. (May 2015)<http://www.cantechletter.com/2015/05/vancouvers-g-kup-ready-to-disrupt-massively-wasteful-single-serve-coffee-pod-market/>.
[43] 2015 Bioplastics Award, Bioplastics Magazine, <http://www.bioplastics-magazine.com/en/events/bioplastics_award.php>.

Index

Note: Page numbers followed by "*f*" and "*t*" refer to figures and tables, respectively.

A

Abiotic oxidation, 31–32
Absorbable anterior and posterior cruciate ligament interference screw, 237*f*
Absorbable orthopedic pin, 237*f*
Acid–base titration, 74
Acrylic-styrene-acrylonitrile (ASA), 254
Additives and modifiers, 153
  bio-based, 8–9
  for bioolefins, 156–160
  for biopolymers, 160–172
    foamability, 167
    heat resistance, 165–167
    mechanical properties, 167–172
    processability, 163–165
    toughness, 160–163
  effect on properties, 172–175
    foamability, 174
    heat resistance, 173–174
    mechanical properties, 175
    processability, 173
    toughness, 172–173
  types of, 154–156
    type 1 additives, 155
    type 2 additives, 155–156
    type 3 additives, 156
Aerobes, 32–33
Aerobic biodegradation, 32–33, 72
Agricultural mulches, 234, 234*f*
Agricultural waste, 252–253
Agriculture and horticulture applications, 234–235
Agro-polymers, 82
Ahlstrom Corp., 202
Aliphatic copolyesters, 83*t*
Aliphatic polyamides, thermal oxidation of, 43, 44*f*
Aliphatic polyester, 36, 83*t*, 104, 112
  general degradation mechanism of, 36, 37*f*
Alkyds, 139–140, 140*f*
Amylopectin, 86, 86*f*
Amylose, 86
Anaerobic biodegradation, 32–33
Arboblend resins, 189
Arkema
  Biostrength, 161, 163
  ski boot, 267
ArmorLite roofing material, 253–254, 253*f*
Aromatic copolyesters (PBAT), 83*t*
Aromatic polyesters
  general degradation mechanism of, 35, 35*f*
  thermal degradation mechanism of, 35–36, 36*f*
ASTM D6400, 1–3
ASTM D6866, 1–3, 139
ASTM sub-committee D20-96, 65*t*
AT&T, 128
Automotive applications, 240–246

B

Bacillus megaterium, 104
Bacterial polyesters, 33, 104–111
  polyhydroxyalkanoates (PHAs), 104–108
  polyhydroxybutyrate (PHBs), 108–111

BASF, 85, 270
  Ecovio resins, 168
Bioactive substances, 61
BioBag, 90
Bio-based additives, 8–9
Bio-based bioplastics, 1, 2*f*, 14–15
Bio-based butadiene (bio-BDO), 85
Bio-based monomers, polymers from, 132–139
  1, 3 propanediol and DuPont's Sorona, 134–135
  bio-based unsaturated polyester, 136–139
  lactic acid and poly(lactic acid), 133
  succinic acid, 136
Bio-based plastics, 5, 26–27, 27*f*, 245*f*
  advantages of, 26–27
Bio-based polyamides, 240–243, 267, 267*f*
Bio-based polyesters, 240–243
Bio-based polyethylene (Bio-PE), 7, 127–128, 128*f*, 268
Bio-based polymers, 81, 84–111
  bacterial polyesters, 104–111
  cellulose-based polymers, 92–103
  chitin and chitosan, 103–104
  different ways to produce, 85*f*
  evolvement of, 84*t*
  starch-based polymers, 104–111
Bio-based resins, 139–141
Bio-based unsaturated polyesters (UPEs), 136–139
Bio-BDO. *See* Bio-based butadiene (bio-BDO)
Bioceta, 94–95
Biodegradability, 23, 31, 61
  example of, 62*f*
  factors affecting, 77–78
  measurement of, 70–77
    gas-evolution test system, 76–77
    plate test system, 73
    respirometric system, 74–76
  standards, 63–67
    levels of standardization, 67*f*
    specifications, 67*f*
    standardization nomenclature, 64*t*
    terminology used in, 65*t*
  testing, 67–70, 68*f*
    field test, 70
    laboratory test, 68, 70
    simulation tests, 70
Biodegradability testing, 67–70
Biodegradable bags, 230–231, 232*f*
Biodegradable bioplastics, 1, 2*f*
Biodegradable biopolymers, 14–15
Biodegradable burial pods, 237*f*
Biodegradable dental implants, 235–238
Biodegradable hangers, 233*f*
Biodegradable materials, 61–62, 180, 251
Biodegradable natural polymer, 205*f*
Biodegradable plastic staples, 235
Biodegradable plastics, 23–24, 65*t*, 72, 227–228, 229*f*, 254
Biodegradable polymers, 19
  application of, 21*f*
  classification of, 20*f*
  definitions, 22–28
    bio-based plastics, 26–27
    biodegradable plastics, 23–24
    biopolymer, 24–26
    oxo-degradable plastics, 27–28
Biodegradable polymers, processing, 179
  blow molding, 192–196
    stretch blow molding, 194–196
  blown film extrusion, 196–197
  compression molding, 187–189
  fiber spinning, 199–206
    electrospinning, 202–206
    melt spinning, 200–202
  film casting, 189–192
  injection molding, 180–187
    DuPont Biomax TPS, 184

Ecovio IS 1335, 186–187
Metabolix Mirel, 184–186
NatureWorks Ingeo Series, 182–183
Novamont Mater-Bi, 184
TianAn Biologic, 186
thermoforming, 197–199
Biodegradable polymers, types of, 81, 84*f*
bio-based monomers, polymers from, 132–139
1, 3 propanediol and DuPont's Sorona, 134–135
bio-based unsaturated polyester, 136–139
lactic acid and poly(lactic acid), 133
succinic acid, 136
bio-based polymers, 84–111
bacterial polyesters, 104–111
cellulose-based polymers, 92–103
chitin and chitosan, 103–104
starch-based polymers, 86–92
bio-derived polyethylene, 127–131
overview of, 83*t*
paper/substrate coatings, bio-based materials for, 139–141
synthetic biodegradable polymers, 112–127
poly(glycolide)/poly(glycolic acid), 121–123
poly(lactic acid), 114–118
poly(lactide-*co*-glycolide), 125–127
poly(*p*-dioxanone), 123–125
poly(ε-caprolactone), 118–121
Biodegradable synthetic polymer, 205*f*
Biodegradation curve, 71, 71*f*
Biodegradation of polymers, 32*f*
Biodegradation process, 19, 20*f*, 31, 61–62, 65*t*
Bio-derived polyethylene (Bio-PE), 84, 127–131, 128*f*
Bio-Fan, 268–269
BIO-FLEX, 194
Bio-Flex F6510, 195–196
BioFoam, 255, 258, 258*f*
Bio-LDPE. *See* Bio-Low density polyethylene (Bio-LDPE)
Bio-Low density polyethylene (Bio-LDPE), 128
Biomax Strong, 155–156, 160–161, 161*f*
Biomax Strong 100 and 120, 160–161, 163, 164*t*
Biomax Thermal 300, 166
Biomer, 112*t*
Bioolefins, additives and modifiers for, 156–160
Bio-PE. *See* Bio-derived polyethylene (Bio-PE)
Bioplastic touch screen computer casing, 241*f*
Biopol, 112*t*
Biopol B, 107, 108*f*
Biopolyesters, 82–83
Biopolymer commercialization, barriers to, 14–15
Biopolymer Network, 255–257
Biopolymers, 14–15, 24–26
additives and modifiers for, 153, 160–172
foamability, 167
heat resistance, 165–167
mechanical properties, 167–172
processability, 163–165
toughness, 160–163
advantages of, 25–26
challenges to, 15
functions of, 25*t*
structures of, 25
BioPreferred coatings, 141
categories, 140*t*
Biorefinery concept, 9*f*
Biostrength, 155–156, 170
Biostrength 130 acrylic modifier, 161

Biostrength 150 impact modifier, 161–162
Biostrength 280 impact modifier, 162
Biostrength 700 acrylic copolymer, 163
Biostrength Advantage, 163
BioWit-tube, 267–268, 267*f*
Bisphenol-A, 8–10, 46
Blow molding, 192–196
  blowing stage, 193
  ejection, 193
  injection stage, 193
  preinjection stage, 192
  stretch blow molding, 194–196
Blown film extrusion, 196–197, 196*f*
Blow-up-ratio (BUR), 196–197
Braskem, 128–129, 130*f*
Bulk degradation, 34–35, 120*f*, 121
BUR. *See* Blow-up-ratio (BUR)
Burial pods, biodegradable, 237*f*

C

CA. *See* Cellulose acetate (CA)
CAB. *See* Cellulose-acetate butyrate (CAB)
Calcification, 51–52
Camphor, 101
  characteristics of, 101–102
  structure of, 101*f*
CAP. *See* Cellulose-acetate propionate (CAP)
Capa, 254–255
Capa brand polycaprolactone, 254–255
Caprolactam, 39–40
Carbonic ester, 46
Cargill, 85
Cellobiose, 92, 93*f*
Celluloid, 99–103
  manufacture, flow chart of, 100*f*
  thermal degradation of, 102*f*
Cellulose, 92–93, 92*f*, 100*f*
Cellulose acetate (CA), 95–98
  preparation mechanism, 96*f*
  properties of, 97*t*
  structure of, 98*f*
Cellulose esters, 94–95
  melting points of, 95*t*
Cellulose nitrate, 100–101, 101*f*
Cellulose-acetate butyrate (CAB), 99, 99*f*
Cellulose-acetate propionate (CAP), 98–99, 98*f*
Cellulose-based polymers, 92–103
Cellulosic feedstocks, 7–8
CelTiX, 155
CEN, 64, 65*t*
Cereplast, 154
CESA-natur product line, 156
Chitin and chitosan, 103–104, 103*f*
Ciba Speciality Chemicals, 159
Cinnamomum camphora, 101
Citric acid, 92
Clariant, 156, 165
  HYDROCEROL chemical foaming agents, 167, 168*f*
Colorants, 102
Commercial applications of bioplastics, 227
  agriculture and horticulture, 234–235
  automotive, 240–246
  bags, 230–231
  biodegradable plastics, 227–228
  consumer electronics, 240
  disposable housewares, 231–233
  medical devices, 235–239
  packaging, 228–230
Compatibilizer, 91
Compostability, 62, 63*f*, 65*t*
Compostable bags, 227–228
Composting, 61–62, 71–72
Compression molding, 187–189
  arrangement, 188*f*
  of tractor hood and speakers, 190*f*
  process parameters of, 188*f*
Consumer electronics, 240, 242*f*

Conventional extrusion of biopolymers, 213–216
Conventional plastic *versus* Ingeo 3D biopolymer, 262*t*
Conventional plastics and bioplastics, differences between, 15*f*
Coordinate system, 3–4
Corbion Purac, 240, 244–246
Corn starch, 91
Cost feasibility, 4–5
Cotton fibers, 187–188
Current market trends, 10–13
Current material trends, 6–10
Cutley, 231
  compostable, 233*f*
  Natur-Ware bio-based, 233*f*

D

Danimer LLC, 154, 165–166
Degree of substitution (DS), 94, 97, 98*t*
Denimite product, 269*f*
Deoxyribonucleic acid (DNA), chemical structure of, 24*f*
Depolymerization, 31
Derived alkoxy radical, decomposition of, 42*f*
Devolatilization unit, 255
DIN, 65*t*
Direct extrusion, 213–214
Disintegrability, 23
Disposable housewares, 231–233, 233*f*
Dissolvable medical screws promoting bone growth, 237*f*
Dow Chemicals, 170–172
Drop-ins, 6–7, 26
DS. *See* Degree of substitution (DS)
DuPont Biomax TPS, 184
  processing conditions for, 185*t*
DuPont Co., 123
DuPont's Sorona, 134–135
  in flooring applications, 135*f*
  synthesis of, 135*f*
DuPont's Sorona EP PTT polyester, 243
Durabio, 271*f*, 272

E

Ecodear Nanoalloys, 258–259
Eco-Delink, 155
Eco-PLA, 119*t*
Ecovative, 269
Ecovative Myco foam packaging, 269*f*
Ecover, 129, 131*f*
Ecovio IS 1335, 186–187
Ecovio L Foam, 259, 270, 270*f*
Electrospinning, 202–206
  parameters, 204*t*
  setup, 203*f*
Electrospun fibers, 205–207, 206*f*
Elvaloy, 155–156
EMforce Bio, 168–170
EMforce Bio additive, 155
EN 13432, 22–24
Enteric esters, 94
EPI Oxo-Biodegradable bags, 231
EPS. *See* Expanded polystyrene (EPS)
Ester linkages, 33–34, 34*f*, 263–264
Esterase, 33–34
Ethanol, 127
Eureka, 254
European Bioplastics Association, 6
Expanded PLA (EPLA) bead and foams, 255–258
Expanded polystyrene (EPS), 255
Expansorb, 261–263
Expansorb-B universal expander liquid, 263*f*
External deckles, 189–191
Extruder screw, 217–218
Extruders
  components of, 211
  single-screw, 211–212
  twin-screw, 213
Extrusion blow molding, 193–194, 193*f*

Extrusion blow molding (*Continued*)
examples of, 194*f*
Extrusion of biopolymers, 211
conventional extrusion of biopolymers, 213–216
LSLF extrusion, 215–216
extrusion screw designs, 217–223
pulverization screw design, 220–221
reactive extrusion screw design, 221–223
Turbo-Screw design, 218–220
starch extrusion, 216–217

F
Fantastic Beach toys, 231
Feather-g-poly(methyl acrylate) plastic, 265
Fermentation process, 107, 110
Fiber spinning, 199–206
electrospinning, 202–206
melt spinning, 200–202
Field tests, 68*f*, 70
Film casting, 189–192, 191*f*
product examples, 192*f*
First-generation additives, 154
FIS Green 300T, 266–267, 266*f*
Fischer esterification, 94
Fischerwerke, 266–267
FKUR, 194
Foam extrusion process, Turbo-Screw in, 218
Foam packaging, 227–228
Foamability, 167, 174
Food packaging, 228
Ford Motor Co, 1
Fringed-micelle regions, 93

G
Galactic (Belgium), 119*t*
Gas-evolution test system, 76–77
GE Geloy, 253
General Electric, 253
General-purpose screw, 180, 182*f*
Genomatica, 85
G-Kup Coffee, 271, 271*f*
Global markets, 5–6
Global production capacities of bioplastics, 7*f*
by 2018, 10*f*
by material type, 8*f*
Glycolide, 122–123
Glycoprotein, 24
Grab-n-Go, 230
Green Guard, 139
Green Seal, 139
Guided-tissue-regeneration membranes, 235–238, 238*f*

H
HDT. *See* Heat deflection temperature (HDT)
Heat deflection temperature (HDT), 165, 173–174
Heat resistance, 165–167, 173–174
Heated nozzles, 187
High impact polystyrene (HIPS), 198–199
High-shear screws, 180
High-speed energy input (HSEI) TSEs, 215
HIPS. *See* High impact polystyrene (HIPS)
Hot runner systems, 187
HYDROCEROL chemical foaming agents, 167
Hydrolysis, 33–34
of acetals, 55–56
Hyperform
HPN-20E, 157–158, 158*f*

I
Impact modifiers, 172–173
Infrared (IR) radiation, 198–199
Ingeo 3D Series, 259–261
Ingeo 3251D TDS, 180, 182
Inherent biodegradability, 65*t*
Injection blow molding, 192–193
examples of, 194*f*
process, 193*f*

Injection molding, 180–187, 181*f*
DuPont Biomax TPS, 184
Ecovio IS 1335, 186–187
Metabolix Mirel, 184–186
NatureWorks Ingeo Series, 182–183
Novamont Mater-Bi, 184
TianAn Biologic, 186
Injection stretch blow molding (ISBM). *See* Stretch blow molding
InnoCore, 254
Insorb absorbable skin stapler, 237*f*
Invista, 85
IRGACLEAR XT 386, 159, 160*f*
IRGASURF SR 100, 159–160
Iris Industries, 269
ISO 472, 65*t*
Isosorbide-based copolyesters, 9–10

J
Japanese Biodegradable Plastics Society, 65*t*
Johnson & Johnson, 129, 131*f*

K
Kenaf fiber, 187–188
Kickstarter, 269
Knitted cellulosic high-performance fibers, 107, 108*f*
Kuredux PGA data sheet, 123*f*
Kureha Chemical Industries, 123

L
Lacear (Mitsui Toatsu Chemical, Japan), 119*t*
LACTEL absorbable polymers, 263–264, 264*f*
Lactic acid, 115
and poly(lactic acid), 133
Lactron (Kanebo Goshen, Japan), 119*t*
Lacty (Shimadzu Corporation, Japan), 119*t*
Lenzing AG, 202
Life cycle of bioplastics, 3, 3*f*
Low-speed late fusion (LSLF)
extrusion, 215–216
twin-screw extruders, 215
LSLF. *See* Low-speed late fusion (LSLF)

M
Masterbatch, 153–154, 165
Material classification system, 4*f*
Medical devices, 235–239
Melt spinning, 200–202, 201*f*
Melting temperature, 77–78
Mer units, 19, 20*f*
Meredian, 167–168
Metabolix Inc., 9–10, 9*f*, 167–168
Metabolix Mirel, 184–186
processing properties of, 185*t*
MHG Biopolymers, 154, 165–168
Microbial degradation, 31
Millad 3988i, 156–157
Millad NX8000, 156–157
Mineralization, 32, 51–52
Minerals, 155
Minimum renewable carbon content categories (MRCC), 139, 141
Mirel, 9–10, 184–186
processing properties of, 185*t*
Mitsubishi Chemical, 9–10, 272
Mixing and conveying elements, 213*f*
Modifiers, additives and. *See* Additives and modifiers
Monomers, defined, 31
MRCC. *See* Minimum renewable carbon content categories (MRCC)
MSU respirometric systems, 74–76
cumulative measurement respirometric system, 74, 74*f*
direct measurement respirometric system, 74–76, 75*f*
Mushroom packaging, 269
Myco Foam, 269, 269*f*

N
Nanoalloy technology, 258–259
  properties of, 260*t*
National standards, 64–67
Natural biodegradable polymers, 82
Natural polymers, 1, 2*f*, 19, 81–82
Natural resins, 2*f*
NatureWorks, 119*t*, 166
NatureWorks Ingeo Series, 182–183, 270
Natur-Ware cutlery, 231
Nestlé & Tetra Pak, 129
New developments of bioplastics, 251
  applications, 266–272
  materials, 251–264
    Ecodear Nanoalloys, 258–259
    Ecovio L Foam, 259
    expanded PLA (EPLA) bead and foams, 255–258
    Expansorb, 261–263
    Ingeo 3D Series, 259–261
    LACTEL absorbable polymers, 263–264
  processing, 264–265
Nodax, 112*t*
Non-biodegradable biopolymers, 14–15
Nonenteric esters, 94
Nontoxic biodegradable polymer sutures, 235
Nova Institute, 5
Novamont Mater-Bi, 184
  processing conditions for, 184*t*
Novamont resins Inc., 90
Novozymes, 85
Nucleic acids, 24
  function of, 25*t*
Nylon, types of, 38*f*

O
OctoDEX, 239, 254
Octoplus, 239, 254
OnCap Bio, 170, 171*f*
OnCap BIO Antifog, 155–156
OPS. *See* Oriented polystyrene (OPS)
Oriented polystyrene (OPS), 198–199
Oxo-Biodegradable bags, 231
Oxo-degradable additives, 27
Oxo-degradable plastics, 27–28

P
Packaging applications, 11–12, 11*f*, 63, 228–230
Pantene, 130–131
Paper/substrate coatings, bio-based materials for, 139–141
Paraloid BPM-500, 170–172, 171*f*
Parietex ProGrip™ laparoscopic self-fixating mesh, 239, 239*f*
Parison, 193–194
Patented additives, 252–253
PBS. *See* Polybutylene succinate (PBS)
PBSA. *See* Polybutylene succinate adipate (PBSA)
PCL. *See* Poly(ε-caprolactone) (PCL)
PDO. *See* 1, 3 Propanediol (PDO)
PE. *See* Polyethylene (PE)
PEA. *See* Polyesteramide (PEA)
Pebax RNew 80R53, 267
Percent mineralization, 72
Perstorp, 254
PET. *See* Polyethylene terephthalate (PET)
Petroplastics versus bioplastics, 5*f*
PGA. *See* Poly(glycolide)/poly (glycolic acid) (PGA)
PHA-based polymeric modifier, 9–10, 9*f*
PHAs. *See* Polyhydroxyalkanoates (PHAs)
PHB. *See* Polyhydroxybutyrate (PHB)
PHBV. *See* Polyhydroxybutyrate-valerate (PHBV)
Photolysis, 42, 54
PLA. *See* Polylactic acid (PLA)
Plantura, 244–246

Plastic sutures, 235
Plasticizers, 8–9, 91–92, 175, 217
Plate test system, 73
PLGA. *See* Poly(lactide-*co*-glycolide) (PLGA)
Pliable braided sutures, 235
Polestar 200P, 155
Poly(ε-caprolactone) (PCL), 31–32, 83*t*, 90, 118–121
  applications of, 122*f*
  -based bioplastics, 254
  structure of, 119*f*
  synthesis of, 120*f*
Poly(diphenolic acid-isophthaloyl chloride), 35–36
Poly(glycolide)/poly(glycolic acid) (PGA), 112, 121–123, 124*f*
  applications of, 124*f*
  structure of, 122*f*
Poly(lactide-*co*-glycolide) (PLGA), 125–127, 126*f*
Poly(*p*-dioxanone) (PDS), 123–125
  commercial products of, 125*t*
  structure of, 124*f*
Poly(propylene succinate) (PPSu), 36
Polyacetals, 53–56
  depolymerization reaction of, 54*f*
  depolymerization under UV radiation, 55*f*
  hydrolysis of, 55–56
  photo-oxidation of, 54, 55*f*
  thermal oxidation of, 53
PolyActive, 239, 254
Polyamide 4,6 (PA-4,6), 38
Polyamide 6 (PA-6), 38
  cis-elimination reaction in, 40
  degradation of, 40*f*
  monomer formation mechanisms in, 39*f*
Polyamide 6,6 (PA-6,6), 38
Polyamide 11 (PA-11), 38
Polyamides, 38–43
  photo-oxidation, 41–43
  thermal degradation, 39–41, 41*t*
  thermal oxidation, 43
Polybutylene succinate (PBS), 91
Polybutylene succinate adipate (PBSA), 91
Polycarbonates, degradation of, 44–51
  Davis and Golden mechanism for, 44, 45*f*
  Kuroda's mechanism for, 47*f*
  Lee's degradation mechanism, 46, 46*f*
  McNeill and Rincon's mechanism for, 47, 49*f*
  Montaudo and Puglisi's mechanism for, 50*f*
Polyesteramide (PEA), 83*t*, 254
Polyesters, 33–38
  hydrolytic degradation mechanism of, 52*f*
  polycondensation and hydrolysis of, 33, 34*f*
  structure of, 34*f*
Polyethylene (PE), 11–12, 127
  bio-derived, 127–131
Polyethylene terephthalate (PET), 11–12, 31–33, 114–115, 244
Polyhydroxyalkanoates (PHAs), 33–34, 83*t*, 104–108
  applications of, 109*f*
  function of, 25*t*
  structure of, 105*f*
  typical properties of, 108*t*
Polyhydroxybutyrate (PHB), 108–111
  commercial products, 112*t*
  life cycle of, 111*f*
Polyhydroxybutyrate-valerate (PHBV), 166–167, 186
Polyisocyanates, 51, 52*f*
Polyketals, 252, 252*f*
Polylactic acid (PLA), 7, 31–32, 81, 83*t*, 112, 114–118, 154, 161–166, 165*f*, 180, 182, 197, 213, 227, 255

Polylactic acid (PLA) (*Continued*)
applications of, 115*f*
glass transition and melting temperatures, comparison, 116*f*
glass transition temperatures for, 117*f*
main products of, 119*t*
PLA golf tee, 237*f*
PLA/PLGA biopolymers, 261–263
synthesis of, 116*f*, 134*f*
thermal degradation of, 117, 118*f*
Polymer degradation, mechanisms of, 31
modes and corresponding factors of, 33*t*
polyacetals, 53–56
hydrolysis, 55–56
photo-oxidation, 54
thermal oxidation, 53
polyamides, 38–43
photo-oxidation, 41–43
thermal degradation, 39–41
thermal oxidation, 43
polycarbonates, 44–51
polyesters, 33–38
polyurethanes, 51–53
Polymer drying, 180
Polymer matrix characteristics, 239
Polymeric materials, biodegradation of, 23*f*
Polymers, defined, 19, 31
Polyolefins, 156
PolyOne, 166–167, 170
OnCap bioadditives, 171*f*
Polyoxymethylene (POM), 53
degradation mechanisms of, 53
thermal degradation of, 53, 54*f*
Polyphenols, function of, 25*t*
Polyphosphates, function of, 25*t*
Polypropylene (PP), 157–159
Polypropylene/polylactic acid (PP/PLA), 244
Polysaccharides, function of, 25*t*
Polysulfates, function of, 25*t*
Polyurethanes (PU), 51–53, 51*f*
POM. *See* Polyoxymethylene (POM)
PPSu. *See* Poly(propylene succinate) (PPSu)
Processability, 163–165, 173
Processing aids, 153
Processing of biopolymers, 211
Processing temperature profile, 183*t*
1, 3 Propanediol (PDO), 134–135
Proteins, function of, 25*t*
Proton-initiated degradation, 41–42
Pseudomonas, 33–34, 104
Pulverization screw design, 220–221
Purac, 257
Pyranose rings, structure of, 93*f*

Q

QMILK, 266, 266*f*

R

Reactive extrusion, 221–222
two-stage screw design for, 223*f*
types of, 223*t*
Reactive extrusion screw design, 221–223
Recycled chicken feathers, 265, 265*f*
Recycled polymers, 252–253
Recycled wood by-product, 252–253, 253*f*
Reedy International, 167
Regional standards, 64–67
Renewable polymers, 82, 85
Resin synthesis, bio-based building blocks, 141*t*
Respirometric system, 74–76
MSU respirometric systems, 74–76
UMIC Lab I, 76
Ribonucleic acid (RNA), chemical structure of, 24*f*
Röchling automotive, 244–246, 246*f*

S

Safoam RPC-20MS1 and 20MS2, 167
Scarpa, 267, 267*f*

Screw flight circulation channel geometry, 219*f*
Seeding strips, 234, 234*f*
Semicrystalline polyesters, 34
Simulation tests, 68*f*, 70
Single-screw extruders, 211–212, 212*f*, 216
Single-stage stretch blow molding, 195
Ski boot, 267
SMR. *See* Sulzer Mixer Reactor (SMR)
Solanyl (Rodenburg Biopolymers, Netherlands), 119*t*
Solaplast compostable films, 236*f*
Solaplast mulching films, 234–235
Solaplast resins, 228–230, 230*f*
Solid-state shear pulverization (SSSP), 220–221, 221*f*, 222*f*
Sorona EP PTT polyester, 243
Spandex, 85
Spin pack, 201, 202*f*
Spinning temperature, 200
Spunbond Ingeo PLA fiber, 203*f*
SSSP. *See* Solid-state shear pulverization (SSSP)
Stabilizers, 153
  for translucent celluloid, 102
Standardization procedure, tests under, 68*t*
Stannous octoate, 118–119, 125–126
Starch, 83*t*
Starch extrusion, 216–217
  screw designs in, 217*f*
Starch synthetic aliphatic polyester blends, 89–90
Starch-based PCL polymers, 90*t*
Starch-based polymers, 86–92, 227
  starch–PBS/PBSA polyester blends, 91
  starch–PVOH blends, 91–92
  starch synthetic aliphatic polyester blends, 89–90
  thermoplastic starch (TPS), 87–89
Starch-based resins, 7
Starch–PBS/PBSA polyester blends, 91
Starch–PVOH blends, 91–92
Stretch blow molding, 194–196
  single-stage, 195
  two-stage, 195–196, 195*f*
Succinic acid, 136
  applications of, 137*t*
  products of, 136*f*
Sukano PLA im S550, 172
Sulzer Chemtech, 255, 257
Sulzer Mixer Reactor (SMR), 255
Superoxide dismutase, 252
SUPLA, 240
Surface erosion
  of polycaprolactone, 120–121
  of polyesters, 34–35
SurModics, 254
Symmetrical splined screw shafts, 214, 215*f*
SynBiosys, 254
Synbra Technology, 257
Synthetic biodegradable polymers, 112–127, 113*t*
  poly(glycolide)/poly(glycolic acid), 121–123
  poly(lactic acid), 114–118
  poly(lactide-*co*-glycolide), 125–127
  poly(*p*-dioxanone), 123–125
  poly(ε-caprolactone), 118–121
Synthetic polymers, 19, 31–32, 81, 204–205

T

Tapes, 234, 234*f*
Tetra Pak, 129, 269
  milk cartons, 269*f*
Textile fiber and bioplastic, 266, 266*f*
Thermoforming, 197–199
Thermoplastic starch (TPS), 87–89
  formulation of wheat starch-based, 89*t*

TianAn Biologic, 167–168, 186
processing conditions for, 186*t*
Toray, 258–259
Toughness, 160–163, 172–173
Toyota, 244
TSE. *See* Twin-screw extruder (TSE)
Turbo-Screw design, 218–220
Twin-bed desiccant dryers, 180
Twin-screw extruder (TSE), 212*f*, 213
high-speed energy input TSEs, 215
low-speed late fusion TSEs, 215
Twin-screw extrusion, 214
Two-stage stretch blow molding, 195–196, 195*f*
TYRX absorbable antibacterial envelope, 238, 238*f*

U
Ultimate biodegradability, 65*t*
UMIC Lab I, 76
Unsaturated polyesters (UPEs), 136–139
UPE. *See* Unsaturated polyesters (UPEs)
US Department of Agriculture (USDA), 1–3
USDA BioPreferred Program, 139

V
Vehicles containing bio-based plastics, 245*f*
Virgin biopolymers, mechanical properties of, 175
Virgin PLA resins, 167–168

W
Wautersia eutropha, 104
Witasek, 267–268

Y
Yogurt, 270*f*

Z
Zeolite, 255
ZIEHL-ABEGG, 268–269
Bionic Bio-Fan, 268–269
ZinCo GmbH, 268